A00 34812-10

W9-BLD-236

MICROSCOPY HANDBOOKS 27

In Situ Hybridization: a practical guide

F

Royal Microscopical Society MICROSCOPY HANDBOOKS

01 An Introduction to the Optical Microscope (2nd Edn)
03 Specimen Preparation for Transmission Electron Microscopy of Materials
04 Histochemical Protein Staining Methods
05 X-Ray Microanalysis in Electron Microscopy for Biologists
06 Lipid Histochemistry
08 Maintaining and Monitoring the Transmission Electron Microscope
09 Qualitative Polarized Light Microscopy
11 An Introduction to Immunocytochemistry (2nd Edn)
12 Introduction to Scanning Acoustic Microscopy
15 Dictionary of Light Microscopy
16 Light Element Analysis in the Transmission Electron Microscope
17 Colloidal Gold: a New Perspective for Cytochemical Marking
18 Autoradiography: a Comprehensive Overview
19 Introduction to Crystallography (2nd Edn)
20 The Operation of Transmission and Scanning Electron Microscopes
21 Cryopreparation of Thin Biological Specimens for Electron Microscopy
22 An Introduction to Surface Analysis by Electron Spectroscopy
23 Basic Measurement Techniques for Light Microscopy
24 The Preparation of Thin Sections of Rocks, Minerals and Ceramics
25 The Role of Microscopy in Semiconductor Failure Analysis
26 Enzyme Histochemistry
27 *In Situ* Hybridization: a Practical Guide

In Situ Hybridization: a practical guide

A.R. Leitch
Queen Mary and Westfield College, Mile End Road, London
E1 4NS, UK

T. Schwarzacher
John Innes Centre for Plant Science Research, Colney Lane, Norwich
NR4 7UJ, UK

D. Jackson
USDA, Plant Gene Expression Center, 800 Buchanan Street,
Albany, CA 94710, USA

I.J. Leitch
Jodrell Laboratory, Royal Botanic Gardens, Kew, Richmond,
Surrey TW9 3AB, UK

In association with the Royal Microscopical Society

QH
452.8
I61
1994

29851529

6-19-96

© BIOS Scientific Publishers Limited, 1994

All rights reserved. No part of this book may be reproduced or transmitted, in any form or by any means, without permission.

First published in the United Kingdom 1994 by
BIOS Scientific Publishers Limited,
St Thomas House, Becket Street, Oxford OX1 1SJ

A CIP catalogue for this book is available from the British Library.

c1

ISBN 1 872748 48 1

Typeset by Saxon Graphics Ltd, Derby, UK
Printed by Information Press Ltd, Oxford, UK

Preface

In situ hybridization is a powerful technique that enables the visualization of nucleic acid probes on target tissues, cells, nuclei and chromosomes, so that the location of the nucleic acid can be determined as *in vivo*. This book is targeted at scientists wanting to learn the principles and practice of *in situ* hybridization. It may also be useful to the experienced user who wishes to change established methods in order to examine material in new ways.

There are many different types of *in situ* hybridization (such as the methods that use fluorochromes, fluorescence *in situ* hybridization or FISH), and these different approaches are described here. The book begins by outlining the areas where *in situ* hybridization has made an impact. The steps of the *in situ* hybridization reaction are then described chapter by chapter, and the reader is guided through the diverse options available at every step, such as the use of whole mount tissue or spread material, colloidal gold and fluorochrome detection systems, and the use of both electron and light microscopy.

Increasingly, *in situ* hybridization has been used to construct physical maps of chromosomes and to analyse chromosome and genome structure. It has also made an important contribution in determining the spatial and temporal expression of genes. *In situ* hybridization can be used to identify and characterize viral and bacterial sequences, determine sex, localize transformation sequences and analyse neurotransmitter messages. Fundamental biological questions have been answered, and the method has also found an important role in medical diagnosis and plant breeding programmes.

This book arose out of a Royal Society and British Council sponsored programme to China in 1991. The aim was to teach the principles and practice of *in situ* hybridization to the Nanjing Academy of Agricultural Sciences. This book is a consequence of the schedules developed for this programme. The methods transferred readily to China and we are confident that the updated schedules presented here are equally transferable to new laboratories.

Acknowledgements

Many of the schedules and methods described in this book were developed in the Karyobiology Group under the direction of Dr J.S. (Pat) Heslop-Harrison. We are indebted to Pat for his help, encouragement and input throughout this work. We also thank Drs K. Anamthawat-Jónsson and G. Coulton and Mrs M. Shi for their helpful comments. We thank BP, Venture Research International and the AFRC who have supported the research leading to the protocols. Boehringer Mannheim GmbH, Amersham International plc, British Biocell International and Cambio kindly supported the publication of this book.

Safety

Attention to safety aspects is an integral part of all laboratory procedures and both the Health and Safety at Work Act and the COSHH regulations impose legal requirements on those persons planning or carrying out such procedures.

In this and other Handbooks every effort has been made to ensure that the recipes, formulae, and practical procedures are accurate and safe. However, it remains the responsibility of the reader to ensure that the procedures which are followed are carried out in a safe manner and that all necessary COSHH requirements have been looked up and implemented. Any specific safety instructions relating to items of laboratory equipment must also be followed.

Contents

Abbreviations

AAF	2-acetylaminofluorene
AEC	3-amino-9-ethylcarbazol
AMCA	7-amino-4-methyl-coumarin-3-acetic acid
APAAP	alkaline phosphatase–anti-alkaline phosphatase
APES	3-aminopropyltriethoxy-silane
BCIP	5-bromo-4-chloro-3-indolylphosphate
BSA	bovine serum albumin
BrdU	bromodeoxyuridine
CBS	chromatic beam splitter
CCD	charge-coupled device
CISS	chromosomal *in situ* suppression
cM	centimorgan
DAB	diaminobenzidine
DAPI	4', 6-diamidino-2-phenylindole
DEPC	diethyl pyrocarbonate
Dnp	dinitrophenol
dNTP	deoxynucleotide triphosphate
EBV	Epstein–Barr virus
EM	electron microscopy
FISH	fluorescence *in situ* hybridization
FITC	fluorescein isothiocyanate
HRPO	horseradish peroxidase
HTLV-1	human T-leukaemia/lymphoma virus type 1
LINEs	long interspersed repeat sequence
LM	light microscopy
N-Aco-AAF	*N*-acetoxy-2-acetylaminofluorene
NBT	nitroblue tetrazolium
NTP	nucleotide triphosphate
PAP	peroxidase–anti-peroxidase
PBS	phosphate-buffered saline
PCR	polymerase chain reaction
PI	propidium iodide
PRINS	primed *in situ* labelling
RFLP	restriction fragment length polymorphism
SDS	sodium dodecyl sulphate or sodium lauryl sulphate
SINEs	short interspersed repeat sequence
TdT	terminal deoxynucleotidyl transferase
TEM	transmission electron microscopy
T_m	melting temperature
Tnp	trinitrophenol
TRITC	tetramethyl rhodamine isothiocyanate
YAC	yeast artificial chromosome

1 An Introduction to *In Situ* Hybridization

1.1 Outline

In situ hybridization is a powerful method to localize nucleic acid sequences (either DNA or RNA) in the cytoplasm, organelles, chromosomes or nuclei of biological material. *In situ* hybridization differs from the analysis of nucleic acids by Southern or Northern hybridization in that the hybridization signal is localized *in situ* and not on a solid support membrane.

The ability to detect nucleic acids *in situ* enables: (i) construction of physical maps of chromosomes; (ii) analysis of chromosome structure and aberrations; (iii) investigation of the structure, function and evolution of chromosomes and genomes; (iv) determination of the spatial and temporal expression of genes; (v) identification and characterization of viruses, viral sequences and bacteria in tissues; (vi) sex determination; (vii) localization of transformation sequences and oncogenes; and (viii) the analysis of neurotransmitter messages. Fundamental biological questions are answered using *in situ* hybridization and the method is of use in medical research and diagnosis and plant breeding programmes.

Understanding *in situ* hybridization requires knowledge of molecular biology, genetics, immunochemistry and histochemistry. The main steps are outlined in *Figure 1.1* and initially involve the preparation of biological material and the labelling of a nucleic acid sequence to form the probe. Labelling a nucleic acid involves the incorporation of either a radioactive or non-radioactive marker which can be detected. Both probe and material are then denatured to make all nucleic acids single-stranded. Then, under controlled experimental conditions, the single-stranded probe anneals or hybridizes to its complementary single-stranded nucleic acid sequence in the biological material to form a new double-stranded molecule that incorporates the label. Finally the sites of hybridization are detected and visualized; the detection methods depend on the type of label attached to the probe.

1.2 Aims of the book

This book aims to provide a comprehensive account of *in situ* hybridization, including the theoretical basis, methods and uses. Chapter 2 outlines and gives examples of the major classes of DNA and RNA sequences that have been located by *in situ* hybridization. Chapters 3–7 describe the theory and function of the steps in the *in situ* hybridization procedure (see *Figure 1.1*). Chapter 8 presents a detailed working schedule for conducting *in situ* hybridization experiments and includes a troubleshooting section and suggested controls. The future prospects of *in situ* hybridization are outlined in Chapter 9.

There are many ways to prepare material, label nucleic acids and visualize probe hybridization sites. Including all variations is impractical, but in each category methods are described with protocols that are used in our laboratories. Details of the commonly used buffers and sources of chemicals are given in the Appendix.

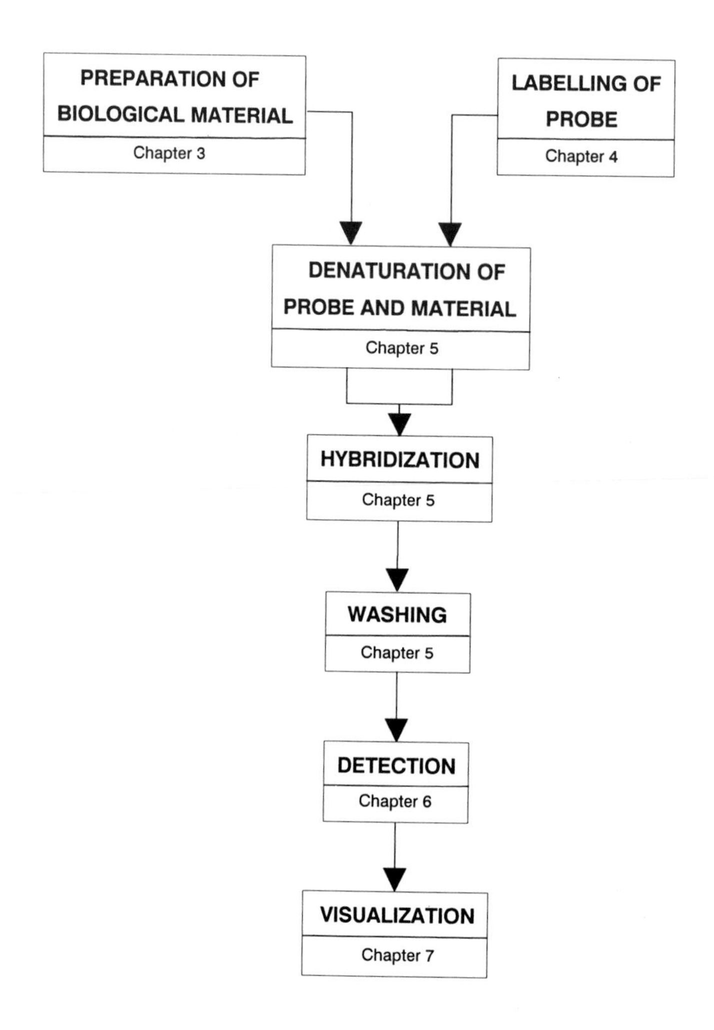

Figure 1.1: An outline of the *in situ* hybridization procedure.

2 Nucleic Acid Sequences Located *In Situ*

2.1 DNA sequences

Many DNA sequences have been visualized by *in situ* hybridization from single-copy to highly repeated sequences. Usually the DNA sequences are hybridized to chromosomes and nuclei prepared from fixed material that has been spread on to glass slides. Depending on the choice of probe, different sequences (target sequences) will be detected (*Table 2.1*).

Table 2.1: Target sequences for probes

Unique or low-copy sequences (Section 2.1.1)

Protein-coding genes
Sequences mapped as restriction fragment length polymorphisms (RFLPs)
Transformation sequences

Repetitive sequences (Section 2.1.2)

Tandem repeats
- Long (9 kbp) ribosomal RNA-coding genes
- Long (5 – 7 kbp) protein-coding genes (e.g. histone-coding genes)
- Short (<300 bp) non-coding units (e.g. sequences at telomeres, heterochromatin, satellite DNA)

Dispersed repeats
- Long interspersed repeats (LINEs, e.g. *Alu*)
- Short interspersed repeats (SINEs, e.g. *Kpn*)
- Retrotransposons and retrotransposon-like sequences

Specific chromosomes or genomes (Section 2.1.3)

mRNA sequences (Section 2.2)

Viral sequences (Section 2.3)

DNA, e.g. EBV (Epstein–Barr virus)
RNA, e.g. HIV (human immunodeficiency virus)

2.1.1 Unique or low-copy sequences

Harper and Saunders (1981) reported the first successful single-copy *in situ* hybridization experiment, localizing the human insulin gene on human chromosome 1 using radioactively labelled probes. *In situ* mapping has now localized, either radioactively or non-radioactively, many unique or low-copy sequences, including genes in human (e.g. Duchenne/Becker muscular dystrophy, Ried *et al.*, 1990; multiple gene mapping, Trask *et al.*, 1993), in *Drosophila* (e.g. the homeotic *Antennapedia* gene complex; Hafen *et al.*, 1983) and in *Caenorhabdites* (e.g. muscle protein genes; Albertson, 1985). Single viral genes can also be mapped on metaphase chromosomes or interphase nuclei (Lawrence *et al.*, 1988). There are only sporadic reports of low-copy *in situ* mapping in plants, although the method has been used to map transformation sequences on plant chromosomes (e.g. T-DNA of *Agrobacterium tumefaciens* introduced into the genome of *Crepis capillaris*; Ambros *et al.*, 1986) and agronomically important genes in cereals (Leitch and Heslop-Harrison, 1993). The detection of single-copy sequences enables the physical location of a gene to be compared with genetic linkage maps and will be increasingly used to analyse genome structure. Multiple labelling and detection systems have now allowed more than one single-copy sequence to be mapped simultaneously with high resolution (Lichter *et al.*, 1990).

Less than 1 kbp of DNA has been detected *in situ* with a signal resolution of about 1 Mbp for metaphase chromosomes (equivalent to about 1 cM, although there is no absolute relationship between physical and genetic distance). Increased resolution of the *in situ* signal to less than 100 kbp is possible using interphase chromosomes, where genes in close proximity may be separately resolved because of the more decondensed chromatin.

The technique of chromosomal *in situ* suppression (CISS) hybridization can facilitate the mapping of cosmid clones containing low-copy sequences by reducing signal from repetitive sequences also present in the clone (Section 4.1.3).

Single- or low-copy sequence detection is perhaps the most demanding and difficult of all *in situ* hybridization experiments. To be certain that low-copy and unique sequences are detected, the *in situ* signal should be present on both chromatids of both homologous chromosomes in a normal diploid cell.

2.1.2 Repetitive sequences

The first sequences detected by *in situ* hybridization were repeated sequences. Gall and Pardue (1969) and John *et al.* (1969) used *in vivo*, radioactively labelled RNA to detect rDNA sequences in cytological preparations. Also amenable to the early *in situ* hybridization experiments were the highly repeated chromosome satellite sequences isolated using buoyant density centrifugation and localized in both mouse and man. Two categories of repeated sequences are described: those in tandem arrays and those dispersed throughout the genome.

Tandem repeats. Repeating units of DNA that are identical, near identical or pattern repeating can occur in tandem arrays of thousands of copies. Tandem repeats include both coding (e.g. histone genes, rRNA genes; *Figure 2.1c* and *d*) and non-coding (e.g. telomeric, *Figure 2.2*; mammalian satellite, *Figure 2.1h*; and plant heterochromatin, *Figure 2.1a*) sequences, many of which have been localized by *in situ* hybridization. Coding sequences tend to have a longer repeat (e.g. the 9 kbp rDNA repeat sequence in plants) than non-coding sequences (e.g. the 150–300 bp satellite sequences that occur around the centromeres of mammalian chromosomes).

In situ localization of tandemly repeated sequences has given considerable information on the structure and evolution of the eukaryote genome. These sequences are also used to detect numerical and structural aberrations of chromosomes, and McFadden (1990) has detected repeat sequences *in situ* to verify endosymbiosis in the evolution of an algal group.

Dispersed repeats. The eukaryote genome also contains repeated sequences that are interspersed among the single-copy sequences. In mammalian genomes the sequences include the short interspersed repeats (SINEs, unit repeat less than 500 bp), long interspersed repeats (LINEs, unit repeats of several kbp) and unclassified spacer sequences. The most abundant human SINEs, the *Alu* sequences (*Figure 2.1g*), are about 300 bp long, constitute about 3% of the total genome and are thought to be RNA pseudogenes. In contrast, the unit repeat of the *Kpn*-LINE family found in primates is between 1.5 and 5 kbp long. *In situ* hybridization shows that LINEs have a different distribution from SINEs and have considerable similarity with retroviral sequences. In cereals, dispersed retrotransposon-like elements have been shown by *in situ* hybridization to be scattered throughout the euchromatic regions of the genome.

2.1.3 Detecting specific chromosomes or genomes

Probes from individual chromosome libraries (e.g. flow-sorted chromosomes cloned into plasmids or cosmids) can be pooled together in a cocktail, labelled and used as a probe. When this probe is hybridized *in situ* with unlabelled blocking (e.g. total genomic) DNA to enhance probe specificity (Section 4.1.3), the entire length of individual chromosomes can be labelled; this is a technique known as 'chromosome painting' (*Figure 2.1e*). Chromosome painting is able to detect numerical and structural chromosome abnormalities. The method has been used for the clinical prenatal and preimplantation diagnosis of chromosome rearrangements (e.g. trisomy 21 responsible for Down syndrome; Lichter *et al.*, 1988). X- and Y-specific probes are also available, which enable the determination of fetal sex and can be used to diagnose sex-linked disorders for which specific genetic tests are not yet available (West, 1990).

Total genomic DNA (genomic probes) with unlabelled blocking DNA (Section 4.1.3) can also be used to identify individual chromosomes in cell fusion hybrids (Manuelidis, 1985; Schardin *et al.*, 1985) and genomes in hybrid organisms (*Figures 2.3* and *2.4a*; Schwarzacher *et al.*, 1989). Genomic probes are increasingly being used in plant breeding to detect alien translocations and substitutions into cereals (*Figure 2.1b* and *f*; Schwarzacher *et al.*, 1992).

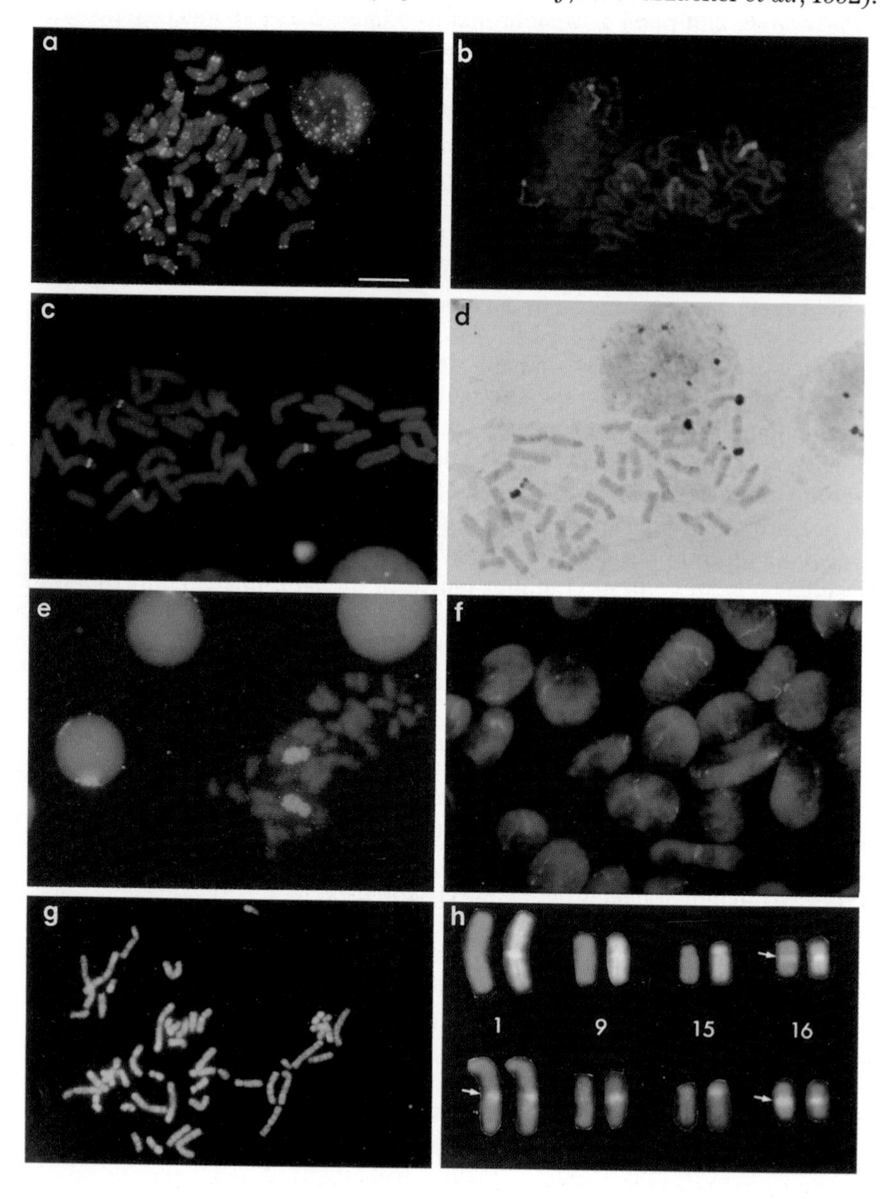

Figure sponsored by Amersham

Figure 2.1: DNA:DNA *in situ* hybridization.

(a) Simultaneous *in situ* hybridization of two differently labelled cloned DNA probes detected by anti-digoxigenin-conjugated fluorescein (green fluorescence) and avidin-conjugated Texas red (orange fluorescence). A root tip chromosome spread of a wheat × rye hybrid (*Triticosecale* cv. Lasko) was probed with a digoxigenin-labelled plasmid pSc119.2 (isolated from rye) and shows the position of the tandem repeat sequence in the heterochromatin (green). Simultaneously, a biotinylated plasmid pTa71 (isolated from wheat) reveals the sites of rDNA (orange). Photograph in collaboration with N. Neves. Scale bar = 15 μm.

(b) Simultaneous *in situ* hybridization of *Thinopyrum bessarabicum* genomic probe, directly labelled with fluorescein, and *Secale cereale* genomic probe, directly labelled with rhodamine. A root tip chromosome spread of a wheat line (*Triticum aestivum* cv. Glennson 1J disomic addition line) containing two chromosomes from the wild grass *T. bessarabicum* and two chromosome arms from rye (*S. cereale*) introduced into the wheat genome using plant breeding methods (kindly provided by T.E. Miller). The *T. bessarabicum* chromosomes fluoresce orange, while the rye chromosome arms fluoresce yellow. Cross-hybridization of probe DNAs to wheat chromosomes was minimized by using 10 times the concentration of unlabelled wheat DNA to probe DNA. Reproduced from *The Chromosome* (1993), p. 178, Figure 2f, BIOS Scientific Publishers. Scale bar = 10 μm.

(c) Fluorescent detection of a digoxigenin-labelled rDNA probe by fluorescein conjugated to anti-digoxigenin. A metaphase spread of wheat (*Triticum aestivum* cv. Chinese Spring) showing sites of rDNA (yellow) following *in situ* hybridization with a ribosomal probe (pTa71). Chromosomes have been counterstained with propidium iodide (orange). Scale bar = 20 μm.

(d) Enzymatic detection of a digoxigenin-labelled rDNA probe by the precipitation of diaminobenzidine catalysed by horseradish peroxidase. A metaphase spread of wheat (*Triticum aestivum* cv. Chinese Spring) showing sites of rDNA (brown precipitate; four major and two minor) after *in situ* hybridization with a ribosomal probe (pTa71). The chromosomes have been counterstained with Giemsa (blue). Scale bar = 20 μm.

(e) Chromosome painting technique showing chromosome 8 in a human metaphase and interphase cell. Pooled clones from a cosmid library of chromosome 8 were labelled with biotin and hybridized in the presence of unlabelled total genomic human DNA in a ratio of 1:1 (= chromosomal *in situ* suppression; see Section 4.1.3). The sites of probe hybridization were detected by fluorescein-conjugated avidin (yellow) and the chromosomes counterstained with propidium iodide (orange). The ability to detect chromosomes at interphase is of great value in determining structural and numerical aberrations in non-dividing cell types. Photograph in collaboration with Professor T. Cremer. Reproduced from *The Chromosome* (1993), p. 178, Figure 2d, BIOS Scientific Publishers. Scale bar = 5 μm.

(f) *In situ* hybridization of digoxigenin-labelled total genomic DNA from rye with unlabelled total genomic DNA from wheat (blocking DNA) to interphase spreads of wheat. The wheat variety carries a translocation between chromosome 1B of wheat and chromosome 1R of rye (1B/1R). Following *in situ* hybridization the rye chromosomes are labelled with fluorescein (yellow) and the wheat chromosomes are counterstained orange with propidium iodide. Total genomic DNA probes can be used to distinguish chromosomes of different parental origin in hybrid plants and show here that the rye chromosomes occur in elongated domains in interphase cells. Photograph in collaboration with Dr R. Kynast. Reproduced from *The Chromosome* (1993), p. 178, Figure 2e, BIOS Scientific Publishers. Scale bar = 10 μm.

(g) *In situ* hybridization of a biotin-labelled synthetic *Alu* consensus sequence (42 bp) detected with avidin-conjugated fluorescein (yellow). The sites of probe hybridization on the human chromosome spread are dispersed across the genome but are absent from some areas, particularly within subcentromeric heterochromatin, which are orange because of the counterstain propidium iodide. Photograph kindly provided by Professor R. Moyzis. Reproduced from Moyzis *et al.* (1989) *Genomics* **4**, 273–289, with kind permission from Academic Press.

(h) Human metaphase chromosomes after *in situ* hybridization using a satellite-related repetitive probe under stringency conditions to give 80–85% sequence similarity (top row) and 60–65% sequence similarity (bottom row). The sites of biotinylated pHuR 195 probe hybridization (arrows) were detected using fluorescein-conjugated avidin and the chromosomes were simultaneously counterstained orange with propidium iodide. Simultaneous staining with 4,6-diamidino-2-phenylindole (DAPI) (blue) shows sites of AT-rich heterochromatin as particularly bright bands. At both stringencies the probe hybridizes to subcentromeric regions on chromosome 16, although the signal strength is higher at low stringency. At low stringency a second hybridization site on chromosome 1 appears. Therefore, the heterochromatin on chromosome 1 shows some sequence similarity with that on chromosome 16. Reproduced from Schwarzacher-Robinson *et al.* (1988) *Cytogenet. Cell Genet.* **47**, 192–196, with permission from S. Karger AG.

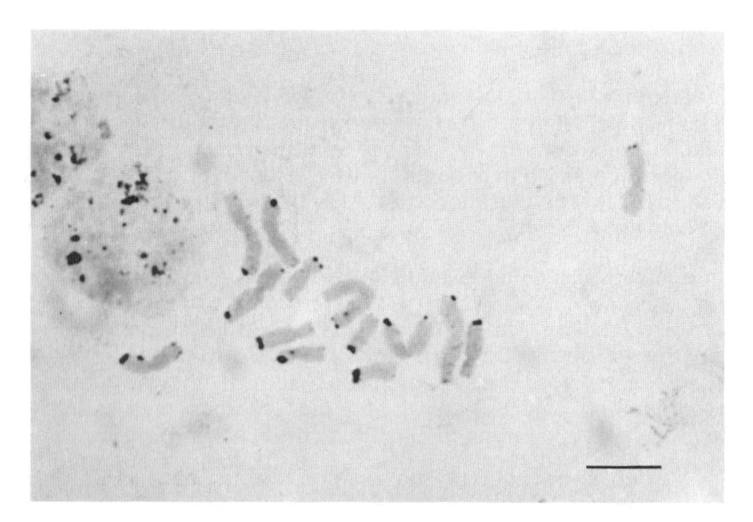

Figure 2.2: Detection of plant telomeres using a biotinylated synthetic oligonucleotide probe by DNA:DNA *in situ* hybridization. The oligonucleotide was derived from the consensus sequence $(TTTAGGG)_n$ from the telomeres of the plant *Arabidopsis thaliana* and end-labelled with biotin-11-dUTP. Sites of probe hybridization to chromosome spreads of *Hordeum vulgare* (barley, 2n = 14) were detected by the avidin–horseradish peroxidase-catalysed precipitation of diaminobenzidine enhanced by silver amplification. The different sizes of the probe hybridization sites may relate to copy number of the target sequence. Scale bar = 10 μm.

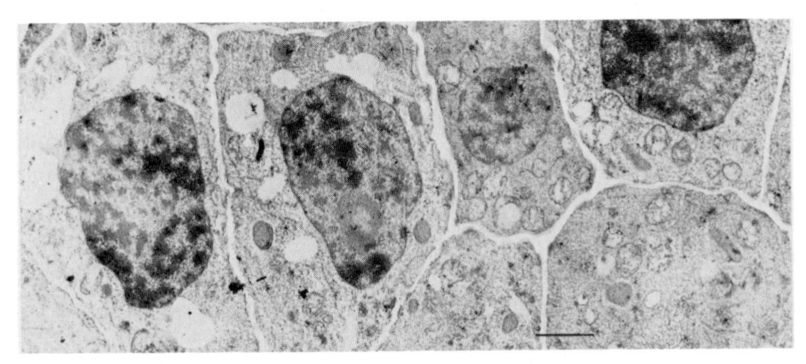

Figure 2.3: Electron microscopic detection of whole genomes on a 0.1 μm resin section through a root tip after DNA:DNA *in situ* hybridization detected by the avidin–horseradish peroxidase precipitation of diaminobenzidine. A section of the hybrid grass *Hordeum chilense* × *Secale africanum* was probed with biotinylated total genomic DNA from *S. africanum*. Chromatin axes of *H. chilense* origin are unlabelled while chromatin axes of *S. africanum* origin are labelled. The section was counterstained with uranyl acetate and lead citrate. Scale bar = 2.5 μm. Reproduced from Leitch *et al.* (1990) with permission from The Company of Biologists Ltd.

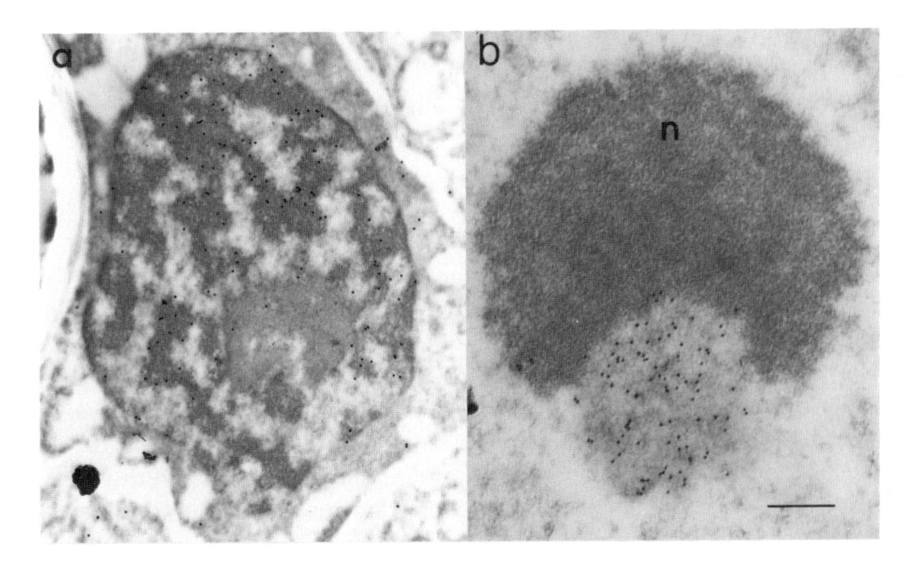

Figure 2.4: Electron microscopic detection of (a) whole genomes and (b) rDNA on 0.1 μm resin sections after DNA:DNA *in situ* hybridization detected by colloidal gold-conjugated avidin (black dots). (a) Sectioned root tip nucleus of *Hordeum chilense* × *Secale africanum* after *in situ* hybridization with biotinylated total genomic DNA from *S. africanum*. Chromatin axes of *S. africanum* origin are labelled with 20 nm gold particles. Scale bar = 1.2 μm. (b) Sectioned root tip nucleus of *Triticum aestivum* (wheat) showing the nucleolus (n) within an interphase nucleus. The section has been hybridized with biotinylated pTa71, which labels rDNA. A large perinucleolar chromatin axis is labelled with 10 nm gold particles. Scale bar = 0.5 μm. Sections were counterstained with uranyl acetate and lead citrate.

Figure sponsored by Biocell.

2.2 RNA sequences

Harrison *et al.* (1974) used tritiated cDNA, synthesized from purified reticulocyte 9S RNA (the globin mRNA), to detect globin gene expression in fetal liver cells. Later, tritiated poly-U probes were used to detect total polyadenylated RNAs, and to optimize many of the experimental parameters for mRNA *in situ* hybridization. Cox *et al.* (1984) were the first to hybridize single-stranded RNA probes prepared by *in vitro* transcription (riboprobes; Section 4.2.3) to detect histone mRNA in sea urchin embryos. Since this major advance a huge number of publications report on the application of riboprobes to both animal and plant studies.

RNA:RNA *in situ* hybridization studies have examined the distribution of RNA sequences in tissue sections or whole-mount specimens. Many experiments have localized mRNA at the intracellular, cellular, tissue and/or organ level to study the spatial and temporal patterns of gene expression. The data can also be analysed semi-quantitatively to estimate the amount of mRNA and correlate mRNA synthesis with protein production, morphogenesis or ultrastructure. In the study of genetic diseases the localization of mRNA transcripts may be more relevant than the localization of the gene itself because the progression of the disease can be monitored in relation to gene activity.

2.2.1 Animal studies

Sea urchin. *In situ* hybridization, in conjunction with immunolocalization studies, has been successfully used in the study of development and differentiation during sea urchin embryogenesis (review: Angerer *et al.*, 1990). Most mRNAs appear to occur in spatially restricted patterns (*Figure 2.5*) with only the abundant maternal mRNAs, expressed during cleavage, being distributed uniformly. Spatially restricted mRNAs probably lead to the specification of cell type fate along both the animal–vegetal and the oral–aboral axes of the embryo. It is possible that most mRNAs are present throughout development although many are subject to spatial regulation. *In situ* hybridization has been used to demonstrate the localization of several tissue-specific mRNAs.

Drosophila. *In situ* hybridization in *Drosophila* developmental studies has helped to establish detailed models explaining how the invertebrate body plan is laid down (review: Ingham *et al.*, 1990). Both maternally and zygotically expressed genes involved in *Drosophila* embryo development have been localized *in situ*. Mutant phenotypes of the genes have contributed significantly to the interpretation of gene expression patterns. The localization of mRNA in both wild-type (*Figure 2.6*) and mutant backgrounds has shown that the restricted patterns of expression are controlled by positive and negative interactions with other gene transcription products.

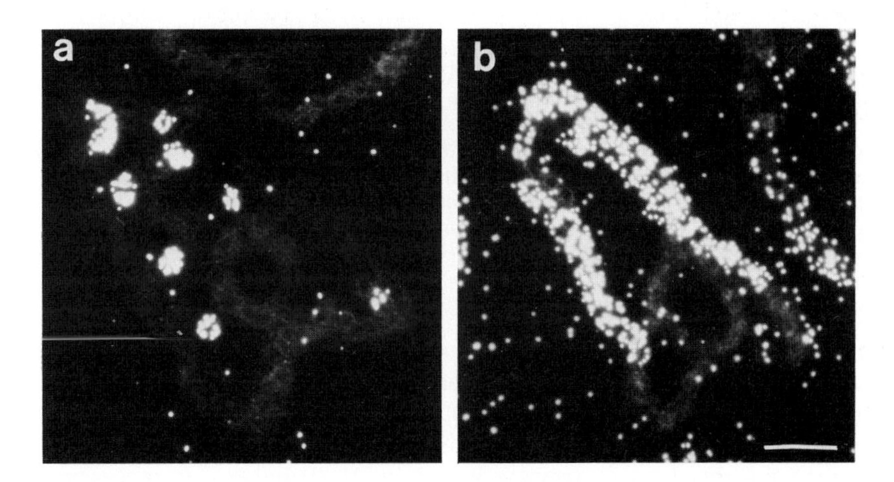

Figure 2.5: RNA:RNA *in situ* hybridization using radioactively labelled riboprobes to (a) primary (skeletogenic) mesenchyme cells and (b) aboral ectoderm cells on sectioned sea urchin embryos. Expression of tissue-specific mRNAs determines cell types before differentiation is observed. The mesenchyme marker indicates the animal–vegetal axis of the embryo, while expression of the aboral ectoderm marker indicates the oral–aboral (dorsal–ventral) axis. Micrographs kindly provided by Drs P.D. Kingsley, L.M. Angerer and R.C. Angerer. Scale bar = 30 μm.

***Xenopus*.** In the study of *Xenopus* oogenesis and embryogenesis, *in situ* hybridization has been used to localize mRNAs (review: Perry O'Keefe *et al.*, 1990). All maternal mRNAs studied in oocytes show a uniform distribution during early oogenesis. As oogenesis proceeds, some mRNAs become localized within the cytoplasm. For example, the maternally encoded gene transcription product, *Vg-1*, is restricted to the vegetal hemisphere in the cortex of the mature oocyte. Following fertilization this localization changes and the transcript becomes distributed in a graded fashion across the vegetal hemisphere. This redistribution is highly suggestive that the *Vg-1* transcription product is required for an early determinative event such as mesoderm induction.

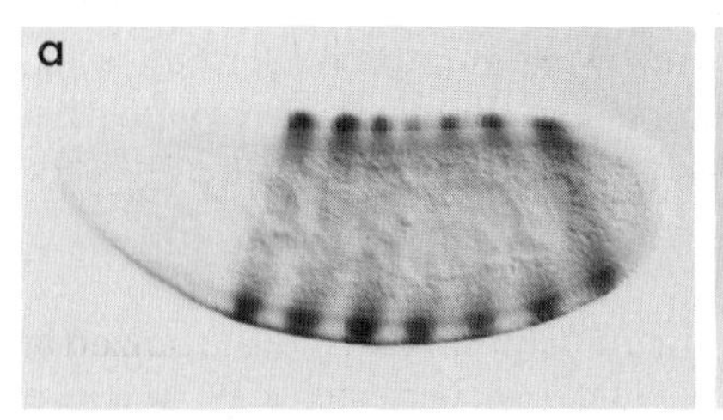

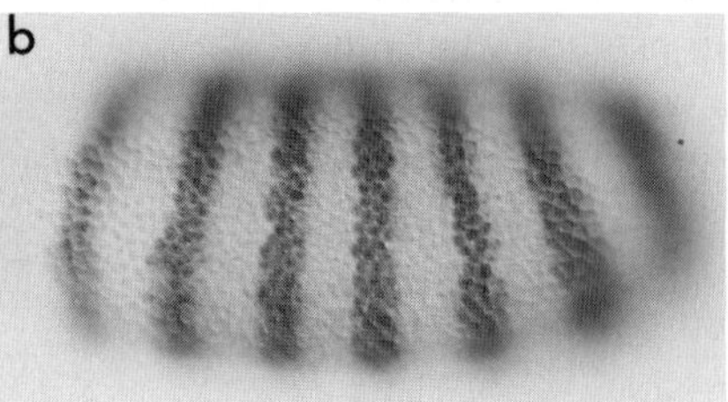

Figure 2.6: RNA:RNA *in situ* hybridization to whole-mount *Drosophila* embryos at (a) low and (b) high resolution showing expression patterns of the segmentation gene *fushi tarazu (ftz)*. The probe was labelled with digoxigenin and detected using the alkaline phosphatase-catalysed precipitation of 5-bromo-3-chloro-3-indolylphosphate/nitroblue tetrazolium (BCIP/NBT). Expression occurs in seven transverse stripes separated by stripes of non-expressing cells in a 2.5 h-old wild-type embryo. The probe demonstrates that the pattern of segmentation becomes established in early embryo development. Micrographs kindly provided by Dr P.W. Ingham.

Mammals. *In situ* hybridization has made a major contribution to the understanding of the role of regulatory peptides in the diffuse neuro-endocrine system by examining the sites of production of mRNAs encoding the peptides (review: Giaid *et al.*, 1990). *In situ* hybridization has also been used to analyse the molecular mechanisms of pattern formation during mammalian embryogenesis (review: Wilkinson, 1990) and an ever-increasing number of regulatory and structural genes have been localized. By using probes from *Drosophila* homeobox genes, homeobox sequences have been successfully identified in mammals. Many experiments have been conducted on the developing central nervous system in the mouse embryo, and these have led to the discovery of several genes (e.g. *int-1*, *Krox-20* and the *Hox* genes) whose expression is spatially regulated. These genes probably play an important role in the early events leading to the formation and spatial organization of tissues.

2.2.2 Plant studies

In plants the regulation of gene expression during development has been shown by *in situ* hybridization. Examples of *in situ* hybridization to plant material include: (i) genes expressed during seed germination; (ii) genes induced around wound sites; (iii) genes involved in the self-incompatibility interaction; (iv) genes regulating floral development; and (v) mRNA distribution during seed storage. Two examples are discussed below.

Floral development. *In situ* hybridization has been combined with genetic analyses to investigate the mechanisms controlling the transition from the vegetative to the floral meristem (*Figure 2.7a* and *b*) and the morphogenesis of the flower (Coen *et al.*, 1990). For example, *in situ* hybridization has shown that the homeotic gene *floricaula* is transiently expressed in a specific and temporal pattern in the bract, sepal, petal and carpel primordia of *Antirrhinum*. The *floricaula* gene product may interact in a sequential manner with other homeotic genes that affect the identity of whorls in the flower.

Seed development. Storage protein gene transcripts have been localized at the cellular level by *in situ* hybridization. In pea cotyledons, expression of the gene encoding the seed storage protein vicilin has been localized to the regions of the embryo which lack mitotic activity. Expression of the gene first occurs in the upper adaxial cotyledon cells and progresses in a wave-like manner through other cells in the cotyledon (Harris *et al.*, 1990).

2.3 Detecting viral sequences

The first viral sequence detected by *in situ* hybridization was the Shope rabbit papillomavirus (Orth *et al.*, 1971). Many types of RNA (e.g. human immunodeficiency virus – HIV) and DNA (e.g. Epstein–Barr virus – EBV, *Figure 2.8*) viral sequences have been localized at the intracellular and tissue levels. Viral sequences may remain separate from the host genome in the cytoplasm (e.g. *Figure 2.8*) or may integrate into one or more chromosomal sites: both types can be distinguished by *in situ* hybridization. *In situ* hybridization enables the simultaneous detection of viral genes and gene transcription products in the same cell preparation, and this has made an important contribution to virology (reviews: Herrington *et al.*, 1990; Teo, 1990).

In situ hybridization has given data on the mode and degree of viral replication, expression and transmission, sites of persistent infection, and the relationship between viruses and disease. Screening diseased tissue by *in situ* hybridization for postulated, uncharacterized viral sequences is a useful approach in the search for covert viral genomes. This approach has suggested an association between multiple sclerosis and a virus possibly related to human T-leukaemia/lymphoma virus type 1 (HTLV-1).

In situ hybridization is not subject to constraints inherent in other methods to detect specific viral sequences. For example, immunological approaches are impossible when antibodies to antigens are unavailable, antigenicity has been lost or viral expression and/or replication has been suppressed. *In situ* hybridization to visualize viral sequences directly has formed the basis of diagnosis and prognosis of viral diseases. However, the sensitivity of *in situ* hybridization is still not sufficient to allow routine detection of viruses integrated into host cells in numbers lower than approximately 10–30 copies.

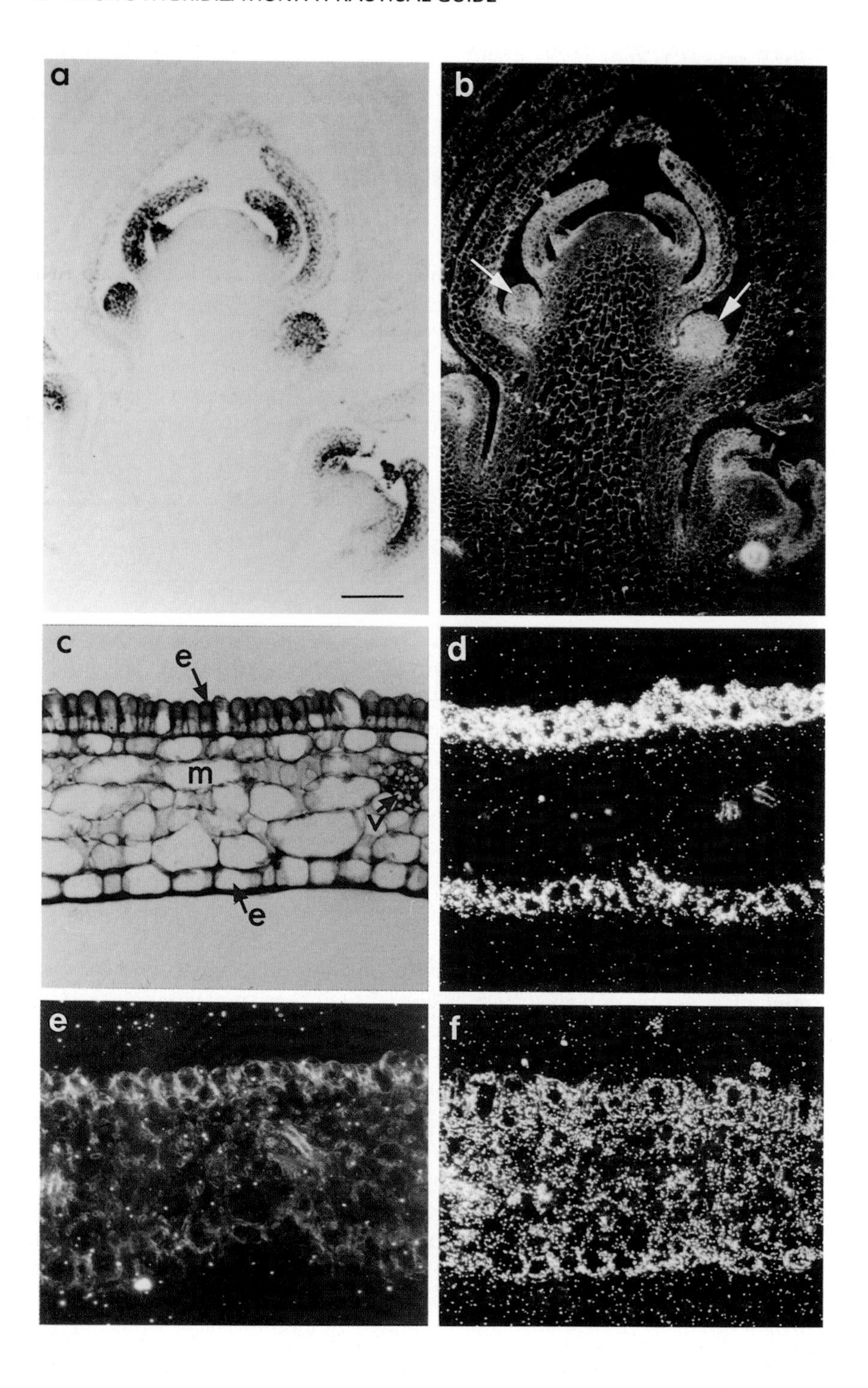
a
b
c
e
m
v
e
d
e
f

◀ **Figure 2.7**: RNA:RNA *in situ* hybridization. (a) and (b) Tissue-specific expression of the *floricaula* gene in an inflorescence shoot that had been embedded in paraffin wax and sectioned. Digoxigenin-labelled *floricaula* probe was detected by the alkaline phosphatase-mediated precipitation of BCIP/NBT and the same section visualized as (a) blue/black precipitate using transmitted light and (b) bright gold/brown regions on an otherwise dark background using reflection contrast. The distribution of *floricaula* changes with the maturity of the apical bracts and is in great abundance in young floral primordia (arrows). Micrographs kindly provided by Drs S. Hantke and R. Coen. Scale bar = 160 μm. (c) to (f) Demonstration of tissue-specific gene expression using ^{3}H-labelled RNA probes. The sites of probe hybridization were detected using autoradiography. Scale bar = 50 μm. (c) Bright-field microscopy of a section through a petal of *Antirrhinum majus* stained with toluidine blue (e, epidermis; m, mesophyll; v, vascular tissue). (d) Section hybridized with the anti-sense chalcone synthase probe. The silver grains are visible as bright spots under dark-field illumination. The expression of chalcone synthase is restricted to the epidermal tissue. (e) Same section as in (d) but hybridized with the sense probe. Note that there is no hybridization above background. (f) Positive control; section hybridized with a ribosomal RNA probe, confirming RNA retention and availability for probe hybridization in all cell types.

Figure sponsored by Boehringer Mannheim.

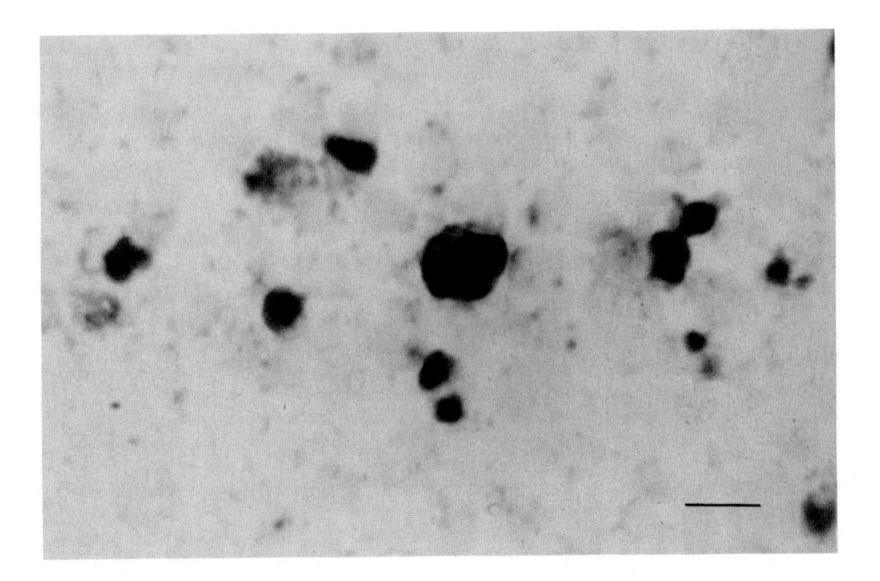

Figure 2.8: Detection of the Epstein–Barr virus (EBV) genome in a marmoset lymphocyte suspension B95-8 culture using DNA:DNA *in situ* hybridization. Cells were cytospun on to a glass slide and hybridized *in situ* with a biotinylated probe containing the *Bam*HI W fragment of the EBV genome. Sites of probe hybridization were localized within some nuclei using the streptavidin–alkaline phosphatase-catalysed precipitation of BCIP/NBT. Many nuclei contain several hundred copies of the EBV genome and are heavily labelled (appearing black), while others show nearly no label. Micrograph kindly provided by Dr L. Labrecque. Scale bar = 2.5 μm.

Further reading

Albertson DG. (1985) Mapping muscle protein genes by *in situ* hybridization using biotin-labelled probes. *EMBO J.* **4**, 2493–2498.

Ambros PF, Matzke MA, Matzke, AJM. (1986) Detection of a 17 kb unique sequence (T-DNA) in plant chromosomes by *in situ* hybridization. *Chromosoma* **94**, 11–18.

Angerer LM, Angerer RC. (1981) Detection of poly A^+ RNA in sea urchin eggs and embryos by quantitative *in situ* hybridization. *Nucleic Acids Res.* **9**, 2819–2840.

Angerer RC, Reynolds SD, Grimwade J, Hurley DL, Yang Q, Kingsley PD, Gagnon ML, Palis J, Angerer LM. (1990) Contributions of the spatial analysis of gene expression to the study of sea urchin development. *Soc. Exp. Biol. Semin. Ser.* **40**, 69–95.

Berleth T, Burri M, Thoma G, Bopp D, Richstein S, Frigerio G, Noll M, Nüsslein-Volhard C. (1988) The role of localization of *bicoid* RNA in organizing the anterior pattern of the *Drosophila* embryo. *EMBO J.* **7**, 1749–1756.

Brandriff B, Gordon L, Trask B. (1991) A new system for high-resolution DNA sequence mapping in interphase pronuclei. *Genomics* **10**, 75–82.

Coen ES, Romero JM, Doyle S, Elliot R, Murphy G, Carpenter R. (1990) *Floricaula*: a homeotic gene required for flower development in *Antirrhinum majus. Cell* **63**, 1311–1322.

Cox KH, Deleon DV, Angerer LM, Angerer RC. (1984) Detection of mRNAs in sea urchin embryos by *in situ* hybridisation using asymmetric RNA probes. *Dev. Biol.* **101**, 485–502.

Ferguson-Smith MA. (1991) Invited editorial: Putting the genetics back into cytogenetics. *Am. J. Hum. Genet.* **48**, 179–182.

Gall JG, Pardue ML. (1969) Formation and detection of RNA–DNA hybrid molecules in cytological preparations. *Genetics* **63**, 378–383.

Giaid A, Gibson SJ, Steel J, Facer P, Polak JM. (1990) The use of complementary RNA probes for the identification and localisation of peptide messenger RNA in the diffuse neuroendocrine system. *Soc. Exp. Biol. Semin. Ser.* **40**, 43–68.

Hafen E, Levine M, Garber RL, Gehring WJ. (1983) An improved *in situ* hybridisation method for the detection of cellular RNAs in *Drosophila* tissue sections and its application for localising transcripts of the homeotic *Antennapedia* gene complex. *EMBO J.* **2**, 617–623.

Harper ME, Saunders GF. (1981) Localization of single copy DNA sequences on G-banded human chromosomes by *in situ* hybridization. *Chromosoma* **83**, 431–439.

Harris N, Mulcrone J, Grindley H. (1990) Tissue preparation techniques for *in situ* hybridization studies of storage-protein gene expression during pea seed development. *Soc. Exp. Biol. Semin. Ser.* **40**, 175–188.

Harrison PR, Conkie D, Affara N, Paul J. (1974) *In situ* localization of globin messenger RNA formation I. During mouse fetal liver development. *J. Cell Biol.* **63**, 401–413.

Herrington CS, Burns J, McGee J O'D. (1990) Non-isotopic *in situ* hybridization in human pathology. *Soc. Exp. Biol. Semin. Ser.* **40**, 241–269.

Ingham PW, Howard KR, Ish-Horowicz D. (1985) Transcription pattern of the *Drosophila* segmentation gene *hairy. Nature* **318**, 439–445.

Ingham PW, Hidalgo A, Taylor AM. (1990) Advantages and limitations of *in situ* hybridization as exemplified by the molecular genetic analysis of *Drosophila* development. *Soc. Exp. Biol. Semin. Ser.* **40**, 97–114.

John HA, Birnstiel ML Jones KW. (1969) RNA–DNA hybrids at the cytological level. *Nature* **223**, 582–587.

Landry ML. (1990) Nucleic acid hybridization in viral diagnosis. *Clin. Biochem.* **23**, 267–277.

Lawrence JB, Villnave CA, Singer RH. (1988) Sensitive, high-resolution chromatin and chromosome mapping *in situ*: presence and orientation of two closely integrated copies of EBV in a lymphoma line. *Cell* **52**, 51–61.

Leitch IJ, Heslop-Harrison JS. (1993) Physical mapping of four sites of 5S rDNA sequences and one site of the α-amylase-2 gene in barley (*Hordeum vulgare*). *Genome* **36**, 517–523.

Lichter P, Cremer T, Tang C-J C, Watkins PC, Manuelidis L, Ward DC. (1988) Rapid detection of human chromosome 21 aberrations by *in situ* hybridization. *Proc. Natl. Acad. Sci. USA* **85**, 9664–9668.

Lichter P, Tang CC, Call K, Hermanson G, Evans GA, Housman D, Ward DC. (1990) High resolution mapping of human chromosome 11 by *in situ* hybridization with cosmid clones. *Science* **247**, 64–69.

Lichter P, Boyle AL, Cremer T, Ward DC. (1991) Analysis of genes and chromosomes by nonisotopic *in situ* hybridization. *Genet. Anal. Techniques Applications* **8**, 24–35.

McFadden GI. (1990) Evolution of algal plastids from eukaryotic endosymbionts. *Soc. Exp. Biol. Semin. Ser.* **40**, 143–156.

McNeil JA, Johnson CV, Carter KC, Singer RH, Lawrence JB. (1991) Localizing DNA and RNA within nuclei and chromosomes by fluorescence *in situ* hybridization. *Genet. Anal., Techniques Applications* **8**, 41–58.

Manuelidis L. (1985) Individual interphase chromosome domains revealed by *in situ* hybridization. *Hum. Genet.* **71**, 288–293.

Moore G, Abbo S, Cheung W, Foote T, Gale M, Koebner R, Leitch A, Leitch I, Money T, Stancombe P, Yano M, Flavell R. (1993) Key features of cereal genome organization as revealed by the use of cytosine methylation-sensitive restriction endonucleases. *Genomics* **15**, 472–482.

Mouras A, Negrutiu I, Horth M, Jacobs M. (1989) From repetitive DNA sequences to single copy gene mapping in plant chromosomes by *in situ* hybridization. *Plant Physiol. Biochem.* **27**, 161–168.

Moyzis RK, Torney DC, Meyne J, Buckingham JM, Wu J-R, Burks C, Sirotkin KM, Goad WB. (1989) The distribution of interspersed repetitive DNA sequences in the human genome. *Genomics* **4**, 273–289.

Orth G, Jeanteur P, Croissant O. (1971) Evidence for and localisation of vegetative viral DNA replication by autoradiographic detection of RNA–DNA hybrids in sections of tumours induced by Shope papilloma virus. *Proc. Natl. Acad Sci. USA* **68**, 1876–1880.

Perry-O'Keefe H, Kintner CR, Yisraeli J, Melton DA. (1990). The use of *in situ* hybridization to study the localization of maternal mRNAs during *Xenopus* oogenesis. *Soc. Exp. Biol. Semin. Ser.* **40**, 115–130.

Ried T, Mahler V, Blonden L, van Ommen GJB, Cremer T, Cremer M. (1990) Direct carrier detection by *in situ* suppression hybridization with cosmid clones of the Duchenne/Becker muscular dystrophy locus. *Hum. Genet.* **85**, 581–586.

Schardin M, Cremer T, Hager HD, Lang M. (1985) Specific staining of human chromosomes in Chinese hamster × man hybrid cell lines demonstrates interphase chromosome territories. *Hum. Genet.* **71**, 281–287.

Schwarzacher T, Leitch AR, Bennett MD, Heslop-Harrison JS. (1989) *In situ* localization of parental genomes in a wide hybrid. *Ann. Bot.* **64**, 315–324.

Schwarzacher T, Anamthawat-Jónsson K, Harrison GE, Islam AKMR, Jia JZ, King IP, Leitch AR, Miller TE, Reader SM, Rogers WJ, Shi M, Heslop-Harrison JS. (1992) Genomic *in situ* hybridization to identify alien chromosome segments in wheat. *Theor. Appl. Genet.* **84**, 778–786.

Tautz D, Pfeifle C. (1989) A non-radioactive *in situ* hybridization method for the localization of specific RNAs in *Drosophila* embryos reveals translational control of the segmentation gene *hunchback*. *Chromosoma* **98**, 81–85.

Teo CG. (1990) *In situ* hybridization in virology. pp. 125–148. In *In Situ Hybridization, Principles and Practice* (eds JM Polak, JO'D McGee). Oxford University Press, New York.

Tkachuk DC, Pinkel D, Kuo WL, Weire HU, Gray JW. (1991) Clinical applications of fluorescence *in situ* hybridization. *Genet. Anal. Techniques Applications* **8**, 67–74.

Trask B, Fertitta A, Christensen M, Youngblom J, Bergmann A, Copeland A, de Jong P, Mohrenweiser H, Olsen A, Carrano A, Tynan K. (1993) Fluorescence *in situ* hybridization mapping of human chromosome 19: cytogenetic band location of 540 cosmids and 70 genes or DNA markers. *Genomics* **15**, 133–145.

Venezky DL, Angerer LM, Angerer RC. (1981) Accumulation of histone repeat transcripts in the sea urchin egg pronucleus. *Cell* **24**, 385–391.

West JD. (1990) Sexing the human conceptus by *in situ* hybridization. *Soc. Exp. Biol. Semin. Ser.* **40**, 205–240.

Wilkinson DG. (1990) mRNA *in situ* hybridization and the study of development. pp. 113–124. In *In Situ Hybridization, Principles and Practice* (eds JM Polak, JO'D McGee). Oxford University Press, New York.

3 The Material

Successful *in situ* hybridization experiments require that the material is adequately preserved and that the target sequences and tissue morphology are maintained. In addition, the tissue has to be permeable to probe and detection reagents.

3.1 Tissue fixation

Fresh material should be used where possible. Once harvested the material should be fixed quickly to minimize endogenous nuclease activity and other degradation processes. If necessary, the material should be cut into small pieces (less than 1 mm thick) for even fixation and to minimize fixative penetration time. Vacuum infiltration may be necessary to allow penetration of fixative into tissues that would otherwise simply float on the surface of the fix because of air spaces. Rapid fixation is particularly important when RNA is to be detected because RNA is very sensitive to the degrading activity of RNase.

The fixation step serves to preserve tissue morphology and to minimize loss of nucleic acids. Two categories of fixative are predominantly used: cross-linking fixatives (e.g. glutaraldehyde, formaldehyde) or protein-precipitating fixatives (e.g. ethanol or methanol mixed with acetic acid in the ratio 3:1).

There is an inverse relationship between the strength of tissue fixation and tissue permeability such that overfixed tissue may be well preserved but may be less accessible to the probe and detection reagents. The fixative chosen depends on the material and probe being used, the method of imaging probe hybridization sites (e.g. light or electron microscopy) and the level of sensitivity required.

Tissues fixed by cross-linking fixatives exhibit good morphology and good retention of target sequence; however, tissue permeability is reduced, inhibiting probe and detection reagent penetration. To increase tissue permeability it may be necessary to perform a permeabilization step before the *in situ* hybridization experiment (Section 3.4). The precipitating fixatives do not significantly reduce tissue permeability but tissue morphology can be poor and loss of target nucleic acids can become a problem.

High-quality chromosome preparations are fixed with precipitating fixatives which enable chromosomes to be spread (Section 3.3.1). Material for sectioning, especially for the detection of mRNA and viral nucleic acids, is usually fixed in glutaraldehyde and/or formaldehyde because the precipitating fixatives can result in loss of nucleic acid.

For each new system, the type and duration of the fixation step should be investigated carefully.

3.2 Slides and electron microscopy (EM) grids

Material for *in situ* hybridization should be mounted on glass slides or gold EM grids. Chromosome and nuclei spreads do not adhere well to unclean slides and consequently can be lost during *in situ* hybridization. Dirt and grease on the slides can also contribute to background signal. The preparation of chromic acid washed slides, which are suitable for DNA:DNA *in situ* hybridization, is described in *Table 3.1*. Sectioned material for light microscopy should be mounted onto slides coated with either poly-L-lysine or activated 3-aminopropyltriethoxy-silane (APES, *Table 3.2a* and *b* respectively). Both reagents coat the slide in charged groups which help bind sections to the slide. For RNA:RNA *in situ* hybridization special precautions must be taken to remove RNase (see Section 8.1). *Table 3.2* includes precautions against RNase activity (i.e. baked coverslips and slides), which are unnecessary steps for DNA:DNA *in situ* hybridization. For electron microscopy, sectioned material is picked up on gold mesh grids (e.g. 400 hexagons) coated with a pyroxylene film (*Table 3.3*).

Table 3.1: Preparation of chromic acid cleaned slides for DNA:DNA *in situ* hybridization

1. Place slides into chromium trioxide solution in 80% (w/v) sulphuric acid for at least 3 h at room temperature. Standard safety precautions for strong acids apply.
2. Wash slides in running water for 5 min.
3. Rinse slides thoroughly in distilled water.
4. Air dry.
5. Place slides into 100% ethanol. Remove and dry slides immediately prior to use.

Table 3.2: Coated slides for sections (suitable for RNA:RNA *in situ* hybridization)

(a) *Poly-L-lysine-coated slides*
1. Place slides in concentrated nitric acid for 30 min.
2. Rinse in deionized water for up to 2 h then air dry.
3. Soak slides in acetone for 15 min and then bake at 180°C for 2 h.
4. After cooling, add a small drop (8 μl) of poly-L-lysine (molecular weight = 300 000, 1 mg ml^{-1} in RNase-free deionized water, see Section 8.1) to each slide and, using a baked coverslip, draw out into a film to cover the slide.*
5. Dry slides overnight on a 40°C hotplate.

(b) *3-Aminopropyltriethoxy-silane (APES)-coated slides*
1. Place slides in concentrated nitric acid for 30 min.
2. Rinse in deionized water for up to 2 h then air dry.
3. Soak slides in acetone for 10 min and then bake at 180°C for 2 h.
4. Dip slides into 2% (v/v) APES in acetone.
5. Rinse thoroughly in acetone.
6. Air dry.
7. Activate APES by placing slides into 2.5% (v/v) glutaraldehyde in 1 × phosphate-buffered saline (PBS) (made up with RNase- free water; Section 8.1) for 1 h.
8. Wash slides in RNase-free deionized water (Section 8.1) and air dry.

* Note that if the sections do not adhere well it is advisable to aerosol spray the poly-L-lysine to make an even coat on the slide.

Table 3.3: Preparation of pyroxylene-coated gold mesh grids for *in situ* hybridization imaged by electron microscopy

Reagents
(a) Butvar solution: 0.15% (w/v) Butvar B98 (Taab Laboratories) in chloroform.
(b) Pyroxylene solution: 4% (w/v) pyroxylene in amyl acetate.

Method
1. Place gold mesh grids onto filter paper and drop approximately 100 μl of butvar solution onto each one.
2. Into a large glass dish containing clean distilled or deionized water place one small drop (approximately 50 μl) of the pyroxylene solution on to the water surface. It should spread into a thin, even film.
3. Place the grids onto the film (butvar surface downwards).
4. Pick up the film by placing filter paper onto the pyroxylene film and picking up the filter paper immediately it becomes wet. The grids and film should become attached to the filter paper.
5. Allow the grids to dry and carbon coat if grid stability is required (this is not usually necessary).
6. Pick up sectioned material on the film side of the grid.

3.3 Preparing material

3.3.1 Cell spreading

The highest quality chromosome preparations are needed for DNA:DNA *in situ* hybridization. In particular, remains of cytoplasm and other cellular material will severely reduce the *in situ* hybridization signal and generate high levels of background. Ideally, there should be little or no contact between cells, but the density should not be so low that cells are difficult to find. Good chromosome preparations should appear with high contrast when examined dry on a microscope slide using phase-contrast microscopy (Section 7.2.1).

Unfortunately, the production of good spreads is the least controllable of all the *in situ* hybridization steps. Different spreads may react differently during the *in situ* hybridization such that a poorly spread metaphase may show no signal while an adjacent good spread will show strong signal. Uneven spreading probably accounts for the patchiness of results that is seen on many slides (Section 8.10.3). Finesse of chromosome spreading comes with experience.

Plants. Any plant tissue containing dividing cells can be used. Usually young root tip meristems (lying immediately behind the root cap) are chosen, but other tissues are possible (e.g. developing anthers, endosperm or apical meristems). The metaphase index in any meristem can be increased using ice water or spindle-inhibiting drugs (e.g. colchicine) prior to fixation. During the handling of plant material, clean tools and glassware and good laboratory practice should be exercised or the metaphase index will be very low.

The cell wall of plants presents problems that do not occur with animal cells by limiting probe penetration and causing high levels of background. Plant material is usually digested with enzymes to remove the cell wall after fixation.

Two methods for preparing mitotic chromosome preparations from plant meristems are described in *Table 3.4:* the squashing method and the dropping method. The squashing method usually gives higher quality metaphase spreads but has the drawback that chromosomes prepared by this method are less sensitive to *in situ* hybridization. The most successful low-copy *in situ* hybridization experiments on plant chromosomes have used the dropping method.

Mammals. In principle, all cell suspension cultures containing mitoses can be used, such as short-term blood cultures and suspensions of trypsinized cultures, e.g. fibroblast or epithelial cultures. Often the metaphase index is increased by colcemide.

For prenatal diagnosis or cancer cytogenetics, primary or monolayer cultures are initiated from amniotic fluid or biopsy material. Culturing cells can induce chromosome changes or give selective advantages to certain cells or cell populations. Therefore *in situ* hybridization on uncultured cells is being increasingly employed but has to rely on the analysis of interphase nuclei.

An important additional step is used with animal cells that is not usually applied to plant material. This step, the hypotonic treatment, swells the cells and separates the chromosomes prior to fixation. The spreads are made by dropping the fixed suspension onto a glass slide. As with plant chromosomes, this stage requires a degree of experience, some laboratories reporting success only when a particular practice is carried out (e.g. angle of slide on to which the suspension is dropped).

Table 3.4: Mitotic chromosome preparations from plant meristems

Reagents

(a) Fixative (freshly prepared): 3 parts 100% ethanol or methanol to 1 part glacial acetic acid.

(b) 10 × enzyme buffer (pH 4.8):
40 mM citric acid
60 mM sodium citrate
Dilute 1:10 in water for use.

(c) 2 × enzyme solution:
2% (w/v) cellulase (1.8% (w/v) dry powder from *Aspergillus niger*, Calbiochem, 0.2% (w/v) 'Onozuka' RS).
20% (v/v) pectinase (from *Aspergillus niger*, solution in 40% glycerol, Sigma).
Make up in 1 × enzyme buffer.
Store in aliquots at −20°C.

Methods

(a) Accumulation of metaphases and fixation

1. To accumulate metaphases, excised root tips, buds or other meristems are treated with one of the following:
   (i) Aerated distilled water at 0°C (ice water) for 24 h (this is ideal for cereals).
   (ii) 0.01–0.05% (w/v) colchicine for 3–6 h at room temperature or 16–24 h at 4°C (works for most plant tissues).
   (iii) 2 mM 8-hydroxyquinoline for 1–2 h at room temperature followed by 1–2 h at 4°C (suitable for dicotyledonous plants, particularly those with small chromosomes, e.g. *Arabidopsis*).
2. Material is then instantly fixed in freshly prepared fixative for at least 10 h at room temperature. Some workers transfer material into 100% ethanol at this stage.
3. Fixed material can be stored for up to 3 months at −20°C prior to making chromosome preparations by either the squashing or dropping method.

(b) Chromosome preparations

(i) Squashing (modified from Schwarzacher et al., *1980*)
1. Wash material 3 × 5 min in 1 × enzyme buffer to remove the fixative.
2. Transfer material into 1 × or 2 × enzyme solution and digest wall material at 37°C until the material is soft, but ensuring the morphology remains (usually 1–2 h). Adjust the time and the concentration of enzyme to suit the material; usually the longer the material has been fixed the longer the digestion that is needed.
3. Wash material in 1 × enzyme buffer for at least 15 min.
4. Transfer material into 45% aqueous acetic acid for 1–3 min.
5. Make chromosome preparations in 45% acetic acid on chromic acid cleaned slides (*Table 3.1*). Use only the meristematic tissue by removing as much of the other tissue as possible (i.e. remove the root cap and tease out the cells in the remaining terminal 1–3 mm). Apply a clean coverslip to the material in a minimal amount of liquid (do not use an acid cleaned coverslip to which chromosomes will stick). Gently disperse the material between glass slide and coverslip by tapping the coverslip gently with a needle and squash the cells using a pressure required to restrict the blood to the thumb nail.
6. Place newly spread slide on to dry ice for 5–10 min or immerse into liquid nitrogen, then flick off the coverslip with a razor blade.
7. Allow the slide to air dry. After making the preparations, slides should be screened and only top-quality spread preparations selected (Section 7.2.1); we routinely discard 50% of the slides after spread making. Spread preparations can be stored desiccated for up to 1 month in the fridge or freezer (−20°C).

(ii) Dropping (modified from Ambros et al. *1986)*
1. Wash material 3 × 5 min in 1 × enzyme buffer to remove the fixative. Remove as much of the non-meristematic tissue as possible (i.e. remove root cap and use terminal 3 mm of root tip).
2. Transfer material into 2 × enzyme solution in a 1.5 ml microcentrifuge tube and digest wall material at 37°C until the material is soft and breaks up easily (usually about 2 h). The material is gently dispersed with a pipette. Adjust the time and the concentration of enzyme to suit the material. Usually, the longer the material has been fixed the longer the digestion that is needed.
3. Centrifuge for 3 min at 800 *g*, discard the supernatant and add 1 ml of fresh 1 × enzyme buffer. Resuspend the pellet with a pipette and leave for 1 min.
4. Repeat step 3 twice.
5. Centrifuge at 800 *g* for 3 min, discard the supernatant, add 1 ml of fresh fixative and resuspend the pellet with a pipette.
6. Repeat step 5 twice, then centrifuge for 3 min at 800 *g*, discard the supernatant and resuspend the pellet in 50–100 μl of fresh fixative.
7. Drop 10–20 μl of cell suspension onto a chromic acid cleaned glass slide (*Table 3.1*) from 5 cm height, and blow gently.
8. Allow the slide to air dry. After making the preparations, slides should be screened and only top-quality spread preparations selected (Section 7.2.1); we routinely discard 50% of the slides after spread making. Spread preparations can be stored desiccated for up to 1 month in the fridge or freezer (−20°C).

Bromodeoxyuridine (BrdU) can be added to cell cultures before harvesting to label late-replicating DNA. BrdU enhances chromosome elongation and enables the detection of late-replicating (G) bands using a labelled antibody to BrdU simultaneously with *in situ* mapping (e.g. Lawrence *et al.*, 1990).

A method for preparing mitotic chromosome preparations from mammalian cell cultures is given in *Table 3.5*. The preparation of chromosomes for EM is described by Narayanswami and Hamkalo (1991).

Drosophila. The method to obtain polytene chromosome spreads is described by Langer-Safer *et al.* (1982). Spreads are made using cells obtained from freshly excised and fixed salivary glands squashed on to glass slides.

3.3.2 Tissue sectioning

To localize cellular DNA and RNA or to detect bacteria, viruses and viral sequences, the material is normally sectioned. The material for sectioning can be fixed and embedded in paraffin wax (e.g. *Figure 2.7a* and *b*) or resin (e.g. *Figures 2.3* and *2.4*) or can be rapidly frozen and cryosectioned. The embedding medium must preserve the target sequences and maintain good morphological structure. A greater sensitivity of detection is achieved when cryosections are used, however the morphological preservation is often poorer, material is harder to handle and certain tissues are not amenable to cryosectioning. The choice of the embedding medium depends on the material used and the sensitivity of detection required. Sections are transferred to glass slides for light microscopy or coated gold grids for electron microscopy (see *Tables 3.1* to *3.3*).

Table 3.5: Mitotic chromosome preparations from mammalian cell cultures (modified from Schwarzacher, 1976)

Reagents

(a) Hypotonic solution:
0.075 M potassium chloride *or*
0.8–1.2% citrate solution (e.g. sodium citrate) *or*
1 part culture medium + 3 parts distilled water.

(b) Fixative (freshly prepared): 3 parts 100% ethanol or 100% methanol to 1 part glacial acetic acid.

Method

1. Treat 10–100 ml culture with 0.01% colcemide for 1–2 h at 37°C to accumulate metaphases (adjust time to cell type and species).
2. Transfer cells into a 15 ml glass or polypropylene centrifuge tube and centrifuge for 10 min at 350–500 *g*.
3. Carefully pour off the supernatant fluid. Resuspend the cell pellet with the last drop of supernatant by shaking. Do not use a pipette.
4. Add about 10 ml prewarmed (37°C or room temperature) hypotonic solution and leave to stand for 10–20 min at 37°C or 20–40 min at room temperature. Hypotonic treatment swells the cells and facilitates untangling of the chromosomes and hence spreading. Time and temperature are critical and vary for cell type and species. Extended treatment and higher temperatures increase swelling of cells. Undertreated cells do not spread well; overtreated cells burst too early and chromosomes are lost.
5. Centrifuge at 350–500 *g* for 10 min.
6. Pour off supernatant and shake up pellet.
7. Resuspend in about 10 ml of fixative. Add the first 1 ml of fixative dropwise and shake well after each drop. Leave for 10 min at room temperature.
8. Repeat steps 5–7 two or three times. This cleans the suspension and improves the quality of the spreads.
9. Cells can be stored in the refrigerator at this point for several days.
10. For spreading, centrifuge the suspension (as for step 5) and resuspend in 0.5–1 ml of fixative. Transfer or drop from a height of a few centimetres, one or two drops of well-mixed suspension to a clean slide (*Table 3.1*). Let dry by blowing or shaking. View under the microscope to check cell density, and quality of spreads. If chromosomes are poorly spread repeat steps 5–7.
11. Select top-quality preparations and store slides desiccated at 4°C (or −20°C) if necessary.

Paraffin sections. Paraffin-embedded material can give excellent preservation of tissue structure (*Figure 2.7*) and enables the collection of serial sections. Paraffin sections are routinely used in pathology and are suitable for *in situ* hybridization experiments. Material must first be fixed and dehydrated prior to embedding in paraffin wax. Sections for light microscopy (7–10 μm thick, minimum 2 μm) can then be cut. After sectioning, the wax is removed to allow probe penetration.

A schedule for fixing and embedding material in paraffin wax for the detection of mRNA is given in *Table 3.6*.

Some material does not embed well in wax, giving poor morphological preservation. In these cases, embedding in water-soluble polyethylene glycol (molecular weight 1000 or 1500) may improve tissue morphology although the tissue sections are harder to handle (Harris *et al.*, 1990).

Resin sections. Material for examination at high resolution using either light microscopy (LM) or EM is embedded in acrylic resins following fixation with cross-linking fixatives such as glutaraldehyde (*Figures 2.3* and *2.4*). Following polymerization, ultrathin sections (0.1–0.25 μm thickness) of the material are cut. Acrylic rather than epoxy resins are used because they are hydrophilic and give better access to the probe and detection reagents.

Table 3.6: Fixing and paraffin-embedding material

Precautions against RNase activity
Precautions against RNase activity outlined in Section 8.1 should be taken at all times when handling material for RNA *in situ* hybridization.

Reagents
(a) Fixative (freshly prepared): 4% (w/v) paraformaldehyde in 1 × PBS, pH 7.2 (for preparation see Section 8.1.1).
(b) Graded ethanol series in saline: prepare 30%, 50%, 75%, 85% (v/v) ethanol in 0.85% (w/v) NaCl, 95% (v/v) ethanol in water, and 100% ethanol. De-gas using a vacuum pump and chill prior to use.

Method
1. Fix material at 4°C overnight (plant material) or 30 min (individual animal cells) to 3 h (animal tissue) at room temperature.
   (i) Tissue is dissected and placed immediately into fixative. If the material floats it must be vacuum infiltrated by submerging it under the surface of the fix (e.g. using wire gauze) then applying a vacuum. The material should sink once the vacuum is released. The fixative should then be renewed since volatile components are lost during vacuum treatment.
   (ii) Cell suspensions can be spun down (200 *g* for 5 min) and resuspended in molten 1% (w/v) low melting point agarose in 1 × PBS (Appendix) at 55°C. Once solidified, the agar block should be refixed for 1 h at room temperature and treated like a tissue block.

2. Wash material in 0.85% (w/v) NaCl for 30 min on ice.
3. Dehydrate plant material in the graded ethanol series (30–100% ethanol) for 90 min (plant material) or 30 min (animal material) each on ice, and 100% ethanol overnight at 4°C.
4. Replace 100% ethanol with fresh 100% ethanol and leave for 1 h (plant material) or 30 min (animal material) at room temperature, then place in 1 part 100% ethanol to 1 part Histo-Clear (Data Diagnostics) for 1 h (plant material) or 30 min (animal material) at room temperature followed by 3 × 1 h in 100% Histo-Clear at room temperature.
5. Embed the material in wax by transferring into fresh Histo-Clear and add half the volume of paraffin wax chippings; leave at 40°C overnight. Transfer to molten wax at 60°C and change the molten wax twice a day for 3 days. This can be reduced empirically depending on the tissue. Inadequately embedded material will not section properly.
6. Place the material into flexible plastic moulds containing molten paraffin wax. Float the mould on water to solidify the wax. Nucleic acids within the wax blocks are stable and blocks can be stored at 4°C for months or years.
7. Once the material is ready for sectioning the paraffin wax block is trimmed to give a trapezoid face with the longer of the parallel sides mounted to strike the microtome blade first. Sections are typically cut to 10 μm thickness. Ribbons of serial sections are floated on water on poly-L-lysine-coated slides (*Table 3.2*). The slides are placed on a hotplate at 40°C for a few minutes to allow the sections to expand. The water is then drained off and the slides left on the hotplate overnight to dry.
8. Sections need to be dewaxed by rinsing twice in Histo-Clear for 10 min each.
9. Rehydrate the sections in the graded ethanol series (100% down to 30% ethanol) for 1 min in each followed by an incubation in 1 × PBS (Appendix) for 5 min prior to *in situ* hybridization.

A protocol for embedding material in acrylic resin for *in situ* hybridization is described in *Table 3.7*.

Table 3.7: Embedding and sectioning material in acrylic resin for *in situ* hybridization

Reagents

(a) Fixative buffer (pH 6.9):
A = 0.1 M Na_2HPO_4
B = 0.1 M KH_2PO_4
Mix 3 parts A to 2 parts B.

(b) Fixative:
2% (v/v) glutaraldehyde (EM grade)
0.2% (v/v) saturated aqueous picric acid
Made up in fixative buffer.

Method

1. Material is fixed for 2 h at room temperature in fixative.
2. Wash 2 × 5 min in fixative buffer.
3. Dehydrate in 10%, 20%, 30%, 50%, 70%, 90%, 100%, 100% ethanol, 10 min in each.
4. Embed in LR White resin (medium grade; London Resin Company) by replacing ethanol with LR White in the following ratios: ethanol–LR White 3:1 (30 min), 1:1 (30 min) 1:3 (30 min) then 100% LR White for 2 days changing the solution five times.
5. Polymerize at 65°C for 15 h.
6. The block containing the material is trimmed to give a trapezoid face with the long axis mounted to strike the knife edge first. Sections are floated onto a solution of 1% (v/v) benzyl alcohol in water to re-expand the sections after compression caused by sectioning. If a diamond knife is used care must be taken because the benzyl alcohol can dissolve the knife's cement. Sections are typically cut to 0.1–0.25 μm thickness.

   Sections for glass slides
   Glass slides are treated with poly-L-lysine (*Table 3.2*) or Vectabond (Vector Laboratories) according to the manufacturer's instructions. Vectabond is a very satisfactory method, giving good section survival and low levels of background. Sections are transferred to the slides with a metal loop (or 2 mm-diameter hole EM grid).

   Sections for EM grids
   Sections are picked up on fine gold mesh grids (e.g. 400 mesh) prepared as described in *Table 3.3*.

Cryosections. Cryosectioning is a fast way to obtain sections; material can be frozen, sectioned and hybridized with a nucleic acid probe on the same day.

Material can be fixed either before (Giaid *et al.*, 1990) or after (Cornish *et al.*, 1987) freezing and sectioning. The success of each method appears to depend on the material used. For mRNA detection, the fixation step usually takes place before freezing to inactivate RNase and to reduce the diffusion of the target sequences.

A protocol for cryosectioning material for *in situ* hybridization is described in *Table 3.8*.

Table 3.8: Cryosectioning plant material for *in situ* hybridization and examination by LM

Reagents
(a) Fixative buffer (pH 6.9):
A = 0.1 M Na_2HPO_4
B = 0.1 M KH_2PO_4
Mix 3 parts A to 2 parts B.

(b) Fixative:
2% (v/v) glutaraldehyde (EM grade)
0.2% (v/v) saturated aqueous picric acid
Made up in fixative buffer.

(c) Cryoprotectant and mountant, e.g. O.C.T. compound (Tissue-Tek, Agar Scientific).

Method
1. Material is fixed for 2 h at room temperature in fixative.
2. Wash 2 × 5 min in fixative buffer.
3. Blot dry material as far as possible.
4. Immerse material in an embedding medium for frozen tissue specimens (Tissue-Tek).
5. Freeze at −20°C.
6. Mount frozen tissue block into a cryostat microtome. Cut thick sections (10–20 μm thick) at −14°C. Ensure the steel blade is sharp and correctly aligned.
7. Pick up sections on a coated slide (*Table 3.2*).
8. Allow sections to thaw and air dry before using for *in situ* hybridization.

3.3.3 Whole-mount preparations

For the study of whole organisms (e.g. *Drosophila* embryos; Tautz and Pfeifle, 1989), organs (e.g. young pea embryos; Harris *et al.*, 1990) or undisrupted cells, material up to 1–2 mm in diameter can be fixed using precipitating or cross-linking fixatives and the *in situ* hybridization protocol conducted on the whole material or on fresh vibratome sections (*Figure 2.6*). This approach is becoming increasingly popular, particularly for the detection of mRNA (see Rosen and Beddington, 1993).

Whole-mount material may have probe and detection reagent penetration problems. Plant cells should be pretreated with cell wall-digesting enzymes (e.g. cellulase and pectinase – as described for making chromosome preparations, *Table 3.4*, step b2). If antibodies are used to detect the sites of probe hybridization (e.g. anti-digoxigenin), background is reduced by preabsorbing the antibodies on control material (not used for the experiment) to remove any non-specific antibody activity. The principal advantage of this technique is that the topography of the tissue is retained. In combination with the confocal microscope or deblurred optical sections (Sections 7.2.5 and 7.2.6), this technique can be used to produce three-dimensional information (e.g. *Figure 3.1*).

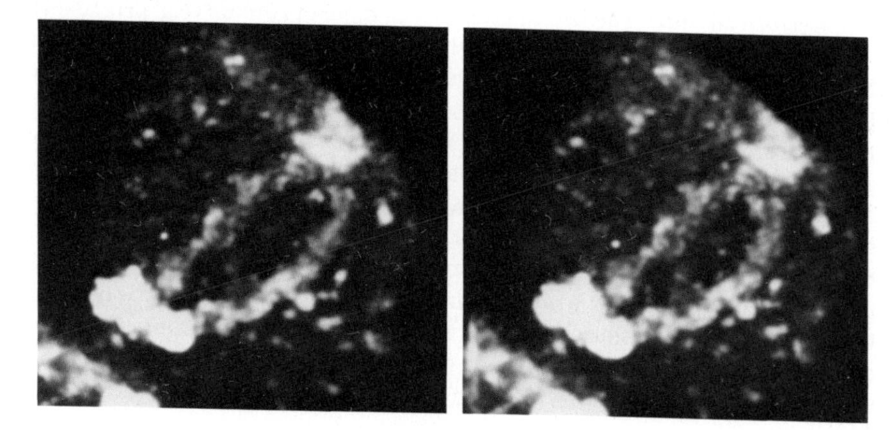

Figure 3.1: A stereo pair of images with 3° rotation showing a whole root tip nucleus after DNA:DNA *in situ* hybridization to detect a pair of chromosome arms. Optical sections were taken with a confocal microscope and the cell reconstructed and displayed on a monitor. The nucleus is from a wheat variety carrying a translocation between chromosome 1B of wheat and chromosome 1R of rye (*Triticum aestivum* cv. Beaver). The rye chromosome arms have been detected using total genomic DNA from rye labelled with biotin and detected with Texas red-conjugated avidin in the presence of unlabelled blocking DNA from wheat. The rye chromosome arms occur in two elongate domains. The large probe hybridization sites at the bottom of the cell arise from subtelomeric heterochromatin which shows little or no decondensation. Photographs taken in collaboration with Dr D. Rawlins.

3.4 Pretreatment of material

Before *in situ* hybridization, material is pretreated to reduce non-specific probe hybridization to non-target nucleic acids and to reduce non-specific interactions with proteins or other components that may bind the probe. These steps also assist probe and detection reagent penetration and maintain the stability of target sequences.

RNase. For the detection of DNA sequences the material is typically treated with RNase A (digests single-stranded RNA) to remove cytoplasmic and nuclear RNA, preventing hybridization with the probe. This is particularly important if the sequence that is being detected is transcribed.

Acetylation. Acetylation neutralizes positively charged molecules, such as basic proteins, and prevents non-specific binding of the probe to the slide when poly-L-lysine-coated slides are used. Acetylation also removes endogenous biotin, which can otherwise cause background signal if a biotinylated probe is used. This pretreatment is optional, although it is often used for RNA:RNA *in situ* hybridization and rarely used for DNA:DNA *in situ* hybridization.

Permeabilization. The use of enzymes to digest proteins (e.g. pronase E, proteinase K or pepsin/HCl) can help with probe and detection reagent accessibility, a process called permeabilization. The enzymes probably act by unmasking nucleic acids from associated proteins; this is particularly necessary when the proteins have been cross-linked with the fixative glutaraldehyde or formaldehyde. This step can be omitted for detecting repeated sequences in cell spreads. When used on cell spread material the concentrations used should be lower than on sectioned material (Section 8.2.1). For example, the range in proteinase K concentration can be from 0.01 $\mu g\ ml^{-1}$ (cell spreads and whole mounts) to 0.5 $\mu g\ ml^{-1}$ (cryosections) and from 1 to 5 $\mu g\ ml^{-1}$ (resin sections). If reagent penetration is problematic the concentrations can be raised, and if cell or tissue morphology is poor the concentration should be lowered.

Prehybridization fixation and material drying. The prehybridization fixation is designed to prevent/reduce diffusion and loss of cellular RNA or DNA. Fixation also stabilizes chromosomes prior to the rigorous denaturation procedures, which can lead to DNA loss.

After fixation the material is dehydrated in an ethanol series and air dried. Although this step is not essential, applying probe to desiccated material ensures that the probe is not diluted by any residual prehybridization solutions. Dehydration is omitted if the material has been embedded in acrylic resin because of adverse reactions of the resin with alcohol.

Protocols for pretreatments are given in Sections 8.1.1 (RNA:RNA *in situ* hybridization) and 8.2.1 (DNA:DNA *in situ* hybridization).

Further reading

Ambros PF, Matzke MA, Matzke AJM. (1986) Detection of a 17 kb unique sequence (T-DNA) in plant chromosomes by *in situ* hybridization. *Chromosoma* **94**, 11–18.

Cornish EC, Pettitt JM, Bonig I, Clarke AE. (1987) Developmentally controlled expression of a gene associated with self incompatibility in *Nicotiana alata. Nature* **326**, 99–102.

Giaid A, Gibson SJ, Steel J, Facer P, Polak JM. (1990) The use of complementary RNA probes for the identification and localisation of peptide messenger RNA in the diffuse neuroendocrine system. *Soc. Exp. Biol. Semin. Ser.* **40**, 43–68.

Harris N, Mulcrone J, Grindley H. (1990) Tissue preparation techniques for *in situ* hybridization studies of storage-protein gene expression during pea seed development. *Soc. Exp. Biol. Semin. Ser.* **40**, 175–188.

Langer-Safer PR, Levine M, Ward DC. (1982) Immunological method for mapping genes on *Drosophila* polytene chromosomes. *Proc. Natl. Acad. Sci. USA* **79**, 4381–4385.

Larsson LI, Hougaard DM. (1990) Optimization of non-radioactive *in situ* hybridization: image analysis of varying pretreatment, hybridization and probe labelling conditions. *Histochemistry* **93**, 347–354.

Lawrence JB, Singer RH, McNeil JA. (1990) Interphase and metaphase resolution of different distances within the human dystrophin gene. *Science* **249**, 928–932.

McFadden GI. (1989) *In situ* hybridization in plants: from macroscopic to ultrastructural resolution. *Cell Biol. Int. Rep.* **13**, 3–21.

McFadden GI, Ahluwalia B, Clarke AE, Fincher GB. (1988) Expression sites and developmental regulation of genes encoding (1–3, 1–4)-β-glucanases in germinated barley. *Planta* **173**, 500–508.

Mouras A, Negrutiu I, Horth M, Jacobs M. (1989) From repetitive DNA sequences to single copy gene mapping in plant chromosomes by *in situ* hybridization. *Plant Physiol. Biochem.* **27**, 161–168.

Narayanswami S, Hamkalo B. (1991) DNA sequence mapping using electron microscopy. *Genet. Anal. Techniq. Applic.* **8**, 14–23.

Rosen B, Beddington RSP. (1993) Whole-mount *in situ* hybridization in the mouse embryo: gene expression in three dimensions. *Trends Genet.* **9**, 162–167.

Schwarzacher HG. (1976) *Chromosomes in Mitosis and Interphase*. p. 182. Springer-Verlag, Berlin.

Schwarzacher T, Ambros P, Schweizer D. (1980) Application of Giemsa banding to orchid karyotype analysis. *Plant Systematics and Evolution* **134**, 293–297.

Tautz D, Pfeifle C. (1989) A non-radioactive *in situ* hybridization method for the localization of specific RNAs in *Drosophila* embryos reveals translational control of the segmentation gene *hunchback*. *Chromosoma* **98**, 81–85.

4 Nucleic Acid Probes, Labels and Labelling Methods

4.1 Probes

DNA sequences are usually detected using labelled DNA probes, while RNA or DNA probes are used for RNA detection. The best probe length for *in situ* hybridization is about 100–300 bases. Shorter probes may result in lower nucleic acid hybrid stability and longer probes (especially >1 kb) may have tissue penetration problems. RNA probes have easier controls than DNA probes, but greater care needs to be taken since RNA probes are more susceptible to degradation by nucleases (*Table 4.1*).

In principle, recombinant DNA technology enables the cloning and purification of any DNA sequence (Section 4.1.1). Alternatively, oligonucleotide sequences can be synthesized *de novo* (Section 4.1.2) or an organism's total genome used (Section 4.1.3). Some of the many ways to label nucleic acids for probes are described in Section 4.2.

4.1.1 Cloned nucleic acids

To amplify a specific DNA sequence by cloning, the DNA is inserted into a vector and both vector and insert are amplified inside appropriate host cells. The amplified DNA is then extracted. Commonly used vectors include bacterial plasmids, bacteriophages (e.g. bacteriophage lambda and M13) and cosmids. Yeast artificial chromosomes (YACs), which will accept up to 1 Mbp of foreign DNA, are increasingly being used for the long-range physical mapping of chromosomes including by *in situ* hybridization (e.g. Baldini *et al.*, 1992). Methods for cloning DNA sequences, maintaining bacterial stocks and isolating DNA and RNA can be found in most standard molecular biology textbooks and papers (e.g. Sambrook *et al.*, 1989) and are not discussed here.

It is advisable to check that cloned sequences conform to expected characteristics prior to any *in situ* hybridization experiment. This is because during cloning inserted sequences can be altered or even lost. The sequence

Table 4.1: Advantages and disadvantages of RNA and DNA probes

	Advantages	Disadvantages
1. *Single-stranded RNA probes (riboprobes)*	Probe free of vector sequences	Sequence must be subcloned into suitable transcription vector
	Post-hybridization RNase A treatment can remove unhybridized single-stranded RNA, thereby reducing non-specific background	RNA is extremely labile, requiring special precautions to be taken
	RNA:RNA hybrid is very stable	RNA probes are sometimes stickier than DNA probes, which may result in a higher background signal
2. *DNA probes*		
(i) *Cloned DNA probes*	The vector sequences may allow for amplification by networking	Probe denaturation required
	No subcloning required	DNA:DNA and DNA:RNA hybrids are less stable than RNA:RNA hybrids
	Easy to handle and store	The double-stranded probe can reanneal to itself during hybridization
(ii) *PCR-generated DNA probes*	Probe free of vector sequences	Sequence must be cloned into a vector containing suitable priming sites unless sequence-specific primers are available
	Sequence amplification and labelling reaction can be carried out simultaneously	DNA:DNA and DNA:RNA hybrids are less stable than RNA:RNA hybrids
		Reaction limited to sequences less than 4 kbp
		Permeability problems may occur with probes longer than 1 kb
(iii) *Synthetic oligonucleotide probes*	Probe can be specifically designed	Oligonucleotide synthesizer required
	Small size of probe (10–50 bp) gives good probe penetration	Small size limits amount of label that can be incorporated
	No cloning required	A few mismatched nucleotide pairs can greatly affect stability of duplex

is usually checked using agarose gel electrophoresis and Southern hybridization experiments.

Double-stranded DNA probes. Cloned DNA sequences are extracted, labelled and used as probes in most *in situ* hybridization experiments (Section 4.2). In some cases the insert is excised from the vector prior to labelling, particularly when the inserted sequence is small in comparison with the vector.

The use of longer probes (such as cosmids or YACs) increases the likelihood that interspersed repetitive sequences are also present (e.g. *Alu* sequences in mammalian chromosomes; Section 2.1.2) which will give rise to additional, unwanted *in situ* signal. This additional signal can be suppressed by including competitive blocking DNA (e.g unlabelled total genomic DNA or the unlabelled Cot-1 DNA fraction), a technique known as chromosomal *in situ* suppression hybridization (CISS; Section 4.1.3; Lichter *et al.*, 1988).

Single-stranded RNA probes. For RNA probes the sequence of interest is inserted into a vector containing transcription initiation sites for bacteriophage RNA polymerases. These sites enable transcription to be initiated *in vitro* in the presence of labelled and unlabelled nucleotides as substrates. The resulting probes are single-stranded RNAs and known as riboprobes (Section 4.2.2).

4.1.2 Synthetic oligonucleotides

Synthetic oligonucleotides are short nucleotide sequences, usually between 10 and 50 bp long, prepared using a DNA synthesizer. They are usually labelled by the end-labelling reaction described in Section 4.2.2. The major advantage of synthetic oligonucleotides is that they can be tailor made to hybridize to specific sequences. Oligonucleotide probes can be used to detect genes, repeat sequences (*Figure 2.1g* and *Figure 2.2*) and RNA.

The short length and low complexity of synthetic oligomers have important implications for the *in situ* hybridization schedule. For example, even a few mismatched pairs will significantly reduce the bond strength of the probe to the target (Section 5.2). Care must therefore be taken in the post-hybridization washes to ensure that hybridized probe is not washed away.

Unlabelled oligonucleotides have been used as primers for directly labelling DNA on chromosomes, a process called primed *in situ* labelling (Section 4.2.5). A similar technique, hybridizing an oligonucleotide probe to mRNA, has also been used to rapidly localize mRNA sequences *in situ* (Section 4.2.5).

4.1.3 Total genomic DNA probes

Total genomic DNA (consisting of the entire DNA complement of an organism's genome) can be labelled and used as a probe (genomic probe) to discriminate genome origins of chromosomes in hybrid plants (*Figures 2.1b* and *f*, *2.3*, and *2.4a*) or to identify individual chromosomes in cell fusion hybrids.

Differentiation between the two genomes is hugely improved when the genomic probe is hybridized in the presence of an excess concentration of unlabelled total genomic DNA from the other genome (=blocking DNA). The blocking DNA hybridizes to common sequences in the probe and on the chromosomes, thereby preventing hybridization of the probe to these sequences. Only sequences specific to the target are available for probe hybridization *in situ*. The blocking DNA may also prevent hybridization of probe DNA to adjacent DNA sequences by steric hindrance. This method exploits the rapid reassociation of highly repeated sequences that are common to the probe and the block. Blocking DNA (total genomic or Cot-1 fraction DNA), used in the technique of CISS, works in a similar way. CISS has been used to improve the specificity of signal when cosmid clones have been hybridized *in situ*, and in the detection of large chromosome segments or individual chromosomes using pooled clones from specific chromosome libraries (chromosome painting; *Figure 2.1e*).

Unlabelled (blocking) genomic DNA also hybridizes to molecules in the cytoplasm and nucleoplasm which could otherwise bind probe or detection reagents, leading to unacceptable levels of non-specific binding. When there is no requirement to block probe hybridization to ubiquitous DNA sequences but it is necessary to reduce non-specific background labelling, salmon sperm DNA is often used.

4.1.4 Polymerase chain reaction-generated DNA probes

The polymerase chain reaction (PCR) has enabled DNA sequences (up to 4 kbp in length) to be specifically and directly amplified. The method can use cloned DNA sequences, using suitable primers that flank the insert (see Section 4.2.2). An important use of PCR amplification methods will be to amplify specific sequences or class of sequences from total genomic DNA or DNA isolated from flow-sorted chromosomal material.

4.2 Nucleic acid labelling

Nucleic acid labels fall into two broad categories, radioactive and non-radioactive. Radioactive and several non-radioactive labels (e.g. biotin, digoxigenin, fluorochromes) can be incorporated into DNA or RNA using enzymatic labelling systems (e.g. for DNA: nick translation, oligolabelling, end labelling; and for RNA: *in vitro* transcription) which synthesize labelled nucleic acids using modified nucleotides (Section 4.2.2). DNA can also be non-radioactively labelled by chemically modifying the DNA helix (e.g. with 2-acetylaminofluorene, mercury; Section 4.2.4). The enzymatic labelling methods usually result in the higher incorporation of modified nucleotides and can therefore generate the most sensitive probes.

For radioactive probes the incorporation of more than one labelled nucleotide can lead to a higher specific activity of the probe but the probe is more expensive to produce and may degrade the probe itself. The incorporation of more than one non-radioactive label in a single probe has not been widely reported but is becoming important in detecting the identity of many different probes simultaneously (Section 6.4; Ried *et al.*, 1992).

Depending on the labelling method, different probe lengths will be generated. For instance, *in vitro* transcription, nick translation and oligolabelling reactions have been optimized to produce a fragmented probe length suitable for *in situ* hybridization. PCR labelling faithfully generates copies of the template DNA which, if greater than about 1 kb, may have cell or tissue penetration problems. Long probes can also be a problem if the nucleic acids are chemically labelled (Section 4.2.4). Treatment of long DNA probes with DNase I reduces probe length.

Hydrolysis steps can be employed to reduce RNA probe length (as described in *Table 4.4*). End-labelling needs to be carefully monitored to optimize incorporation of nucleotides at the end of the probe; if these probes are too long the stability of the nucleic acid hybrid is reduced.

4.2.1 Enzymatic labelling – the labels

Radioactive labels. The radioactive labelling of nucleic acids is usually achieved by enzymatically incorporating nucleotides containing ^{32}P, ^{35}S, ^{125}I or ^{3}H. The labelled nucleotide is essentially identical to its unlabelled counterpart. The labelled probe is not susceptible to steric hindrance during hybridization and this may be one reason why radioactive probes can be more sensitive than non-radioactive probes.

The particular isotope that is chosen will depend upon the application since there is an inverse relationship between sensitivity and resolution (*Table 4.2*). For example, the high emission energy (E_{max}) of ^{32}P makes it suitable for detecting sequences in less than 7 days but gives a wide scattering of silver grains in photographic emulsions during autoradiography (Section 6.1) and low resolution of the signal. In contrast, the weak β-emission of ^{3}H results in a high resolution of signal but it can take at least 2 weeks to expose the photographic emulsion. Isotopes with intermediate energy, such as ^{35}S and ^{125}I, often provide a useful compromise. Ultimately the choice of isotope depends on the particular requirements of the experiment in question.

Non-radioactive labels. In recent years radioisotopes have begun to be replaced by non-radioactive labels. In general, non-radioactive probes may be safer, although toxicity may not be fully known and contamination is less easy to detect than with radioactive probes. In addition, specialized designated laboratories are not needed. Non-radioactively labelled probes can be stored for long periods without loss of activity, the hybridization sites can be detected quickly (hours rather than days or weeks) and several sequences can be detected simultaneously by using different probe labels. Two types of non-radioactive *in situ* hybridization procedures can be distinguished: the direct and the indirect methods.

Table 4.2: Properties of radioisotopes commonly used to label nucleic acids for *in situ* hybridization

Isotope	Particle emitted	E_{max}* (MeV)	Half-life	Resolution† (μm)	Approximate exposure time
^{3}H	β	0.018	12.4 years	0.5–1	2–12 weeks
^{125}I	β γ	0.004 0.035	60 days	1–10	12–20 days
^{35}S	β	0.167	87.4 days	10–15	12–20 days
^{32}P	β	1.71	14.3 days	20–30	2–5 days
^{33}P	β	0.25	28 days	15–20	10–20 days

* Maximum energy of emission.
† Scatter of silver grains in emulsion around point source.

In the direct method the label that has been incorporated into the nucleic acid can be visualized directly once *in situ* hybridization has been completed. For example, Bauman *et al.* (1980) chemically coupled the fluorochrome tetramethyl rhodamine isothiocyanate (TRITC, a rhodamine derivative) onto the 3'-terminus of RNA, which was then hybridized to its complementary DNA sequence *in situ*. The recently introduced fluorochrome-labelled nucleotides (e.g. fluorescein isothiocyanate (FITC), rhodamine (TRITC), 7-amino-4-methyl-coumarin-3-acetic acid (AMCA) linked deoxynucleotide triphosphates (dNTPs)), which can be enzymatically incorporated into the nucleic acid (Section 4.2.2), are becoming extremely important because of the simplicity of detection (*Figure 2.1b*; Wiegant *et al.*, 1991). However, they may not be sensitive enough to detect low- and single-copy sequences. The availability of an anti-fluorescein antibody conjugated to fluorescein enables the signal to be amplified and may increase the sensitivity of detection.

With indirect methods the label that is incorporated into the probe cannot be directly visualized; instead, a second molecule, the reporter molecule, is attached to the probe label after hybridization and enables visualization of the probe hybridization sites.

A range of non-radioactive labels can be incorporated into nucleic acids (*Table 4.3*); the most commonly used are biotin and digoxigenin (described below). The majority of labels can be detected by immunohistochemistry. In the case of biotin either anti-biotin antibodies or (strept)avidin can be used to detect the sites of the probe hybridization. Detection systems for non-radioactively labelled probes are described in Section 6.2.

Table 4.3: Some examples of non-radioactive labels that can be incorporated into DNA and RNA probes

(Photo)biotin
(Photo)digoxigenin
2-Acetylaminofluorene (AAF)
Sulphone groups
Mercury
Bromodeoxyuridine (BrdU)
Fluorochromes
Dinitrophenol (Dnp)

Biotin. Biotin (vitamin H, found abundantly in egg white) was the first non-radioactive label to be used and was incorporated into the nucleic acid in the form of biotin-11-dUTP. Biotin-11-dUTP is the most widely used biotin derivative, but other modified nucleotides are now available, e.g. biotin-16-dUTP, biotin-14-dATP and biotin-11-dCTP (suppliers of biotinylated nucleotides include Enzo Diagnostics, Sigma and Boehringer Mannheim).

Each of the nucleotides is modified at a position that does not interfere with hydrogen bonding between the probe and the target nucleic acid. In addition, each contains a linker arm of at least 11 carbon atoms to ensure access of the detection reagents and to minimize steric hindrance during probe hybridization. Biotin can also be chemically or photochemically (photobiotin; Section 4.2.4) incorporated to make a nucleic acid probe.

Digoxigenin. Digoxigenin is a steroid that occurs naturally in the plants *Digitalis purpura* and *D. lanata*. Digoxigenin is usually incorporated into the DNA or RNA enzymatically by using the nucleotide derivative digoxigenin-11-dUTP or digoxigenin-11-UTP respectively (supplier of digoxigenin-labelled nucleotides is Boehringer Mannheim).

Nucleic acids can also be labelled with digoxigenin photochemically by using the derivative photodigoxigenin (Section 4.2.4).

4.2.2 Enzymatic labelling – the methods

The enzymatic labelling procedures can be divided into those that result in uniformly labelled probes (*in vitro* transcription, nick translation, oligolabelling) and those that label the end of a nucleic acid strand (end-labelling). Uniformly labelled probes are most commonly used as they incorporate more label than end-labelling procedures. However, end-labelled

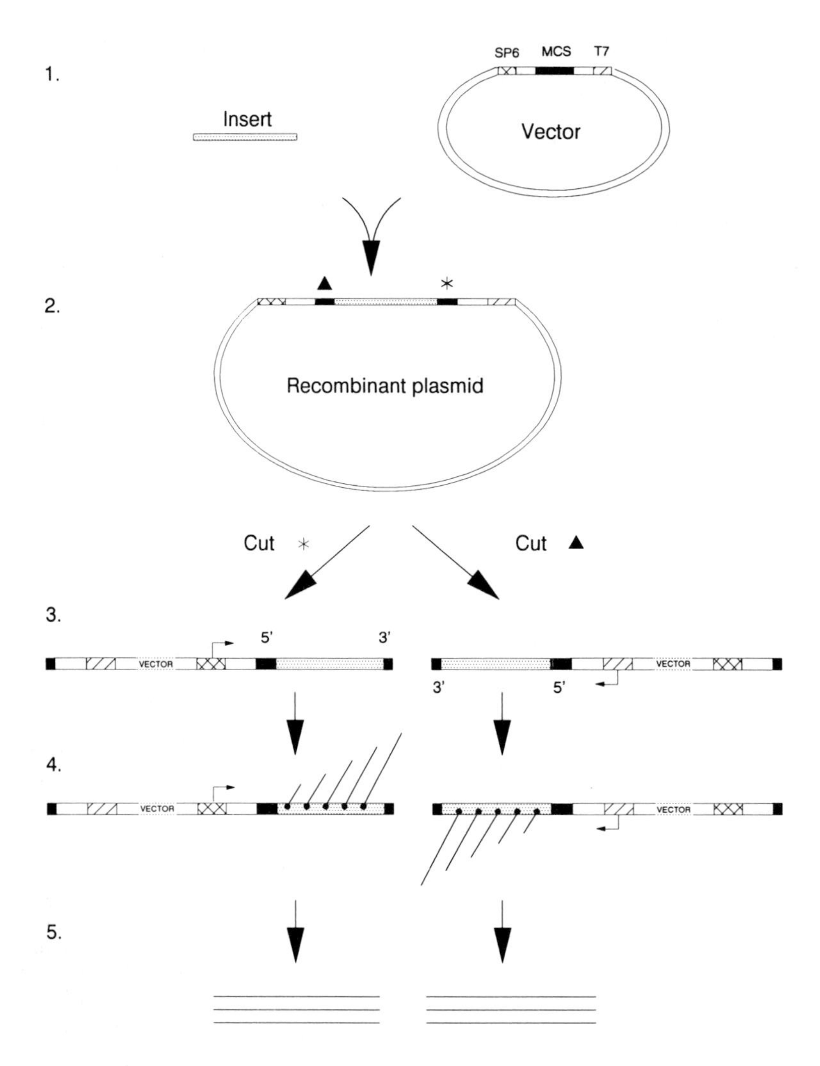

Figure 4.1: The *in vitro* transcription reaction.

1. The reaction utilizes bacteriophage RNA polymerase to synthesize RNA from linearized recombinant plasmids. Cloned DNA is inserted into the multiple cloning site (MCS) of a transcription vector containing an RNA polymerase promoter (e.g. SP6 promoter: cross-hatch; and T7 promoter: hatch) on either side of the multiple cloning site.
2. The recombinant plasmid is linearized using restriction enzymes within the multiple cloning site, downstream of the insert in relation to the chosen promoter (e.g. * for SP6 promoter and triangle for T7 promoter).
3. In the presence of labelled and unlabelled nucleotide triphosphates (NTPs) the RNA polymerase, specific for its promoter, uses the DNA template to synthesize RNA.
4. The 5' to 3' RNA polymerase activity generates single-stranded, labelled RNA transcripts of defined length and sequence.
5. At the end of the reaction the DNA template is removed by digesting with RNase-free DNase.

probes may be more sensitive because of a reduction in steric hindrance between the probe and the target DNA (Cook *et al.*, 1988).

In vitro *transcription – riboprobes.* The sequence of interest is cloned into a vector (e.g. Bluescript from Stratagene; Gemini vectors from Promega) containing the bacteriophage RNA polymerase promoter sequences (e.g. T3, T7 or SP6 RNA polymerase; *Figure 4.1*). In the presence of the appropriate RNA polymerase and labelled (radioactive or non-radioactive labels) and unlabelled nucleotides, single-stranded RNA probes (riboprobes) are synthesized. The high specificity of a bacteriophage polymerase for its promoter enables these vectors to be used to generate single-stranded RNA probes complementary to the coding (sense) or non-coding (anti-sense) strands. For the detection of single-stranded mRNA, the coding strand provides a useful negative control for the specificity of the positive, i.e. anti-sense probe.

Prior to *in vitro* transcription the vector is linearized (*Figure 4.1*, step 2), preferably using a restriction enzyme that generates blunt or protruding 5'-termini.

A protocol for labelling RNA by *in vitro* transcription using tritium (^{3}H) or digoxigenin is given in *Table 4.4*. A kit is available from Amersham International to incorporate fluorescein-11-UTP into RNA by *in vitro* transcription.

Nick translation. The nick translation reaction employs two enzymes, DNase I and *Escherichia coli* DNA polymerase, which incorporate labelled nucleotides along both strands of the DNA duplex (*Figure 4.2*). Usually only one labelled nucleotide is used. The most critical parameter in the reaction is the activity of DNase I, which 'nicks' double-stranded DNA (*Figure 4.2*, step 2); too little nicking can lead to inefficient incorporation of the label and probes that are too long. Too much nicking results in probes that are too short. Commercially available enzyme mixtures of *E. coli* DNA polymerase I and DNase I are now available which are optimized to produce > 50% incorporation of label into DNA in 60–90 min, giving probe lengths of about 200–400 bp.

A protocol for labelling DNA by nick translation is given in *Table 4.5*.

Oligolabelling (random-primed labelling). Oligolabelling of DNA uses the Klenow fragment of *E. coli* DNA polymerase (5' to 3' polymerase activity) and a mixture of all possible oligonucleotides (usually hexanucleotides) that serve as primers for the DNA polymerase (*Figure 4.3*). Since practically all sequence combinations are represented in the hexanucleotide primer mix, the primers bind to the template, on average, every 80–100 bases. The new strand is synthesized by the Klenow fragment of DNA polymerase in a reaction mixture containing labelled and unlabelled nucleotides. The reaction offers the following advantages:

▶ p. 46

Table 4.4: Labelling RNA with [^{3}H]UTP or digoxigenin-11-UTP by *in vitro* transcription

Reagents

(a) 10× reaction buffer for SP6, T7, T3 RNA polymerase:
0.4 M Tris·HCl, pH 8.0 (Appendix)
0.06 M $MgCl_2$
0.1 M dithiothreitol
0.02 M spermidine
0.1 M NaCl.

(b) Unlabelled nucleotide mix: CTP, GTP and ATP individual nucleotide solutions (10 mM); prepare a 1:1:1 mixture.

(c) Labelled nucleotide: *Either*
  (i) 10 μCi of [5, 6-^{3}H]UTP (New England Nuclear, specific activity 35–50 Ci mmol^{-1}. The nucleotide is supplied in 50% ethanol. Remove correct volume containing 10 μCi, dry in a Speedvac and then resuspend in the volume of water required to make the total volume of the reaction mixture 25 μl. *Or*
  (ii) Mix digoxigenin-11-UTP (1 mM stock solution; Boehringer Mannheim) and UTP (1 mM stock) to a final concentration of 0.35 mM digoxigenin-11-UTP and 0.65 mM UTP.

(d) RNase inhibitor: RNAguard ribonuclease inhibitor (Pharmacia) solution = 10 units μl^{-1}.

(e) DNA: Linearize the DNA template (in transcription vector) using appropriate restriction enzyme (see Sambrook *et al.*, 1989) and resuspend in 1× TE (Appendix) to give a final concentration of approximately 1 mg ml^{-1}.

(f) SP6, T7 or T3 RNA polymerase enzyme: The activity of the stock enzyme varies between companies; add the appropriate volume to give a final activity of 0.4 units μl^{-1}.

(g) 2× carbonate buffer, pH 10.2:
80 mM $NaHCO_3$
120 mM Na_2CO_3.

Method

1. Mix the following in a 1.5 ml microfuge tube:
   3 μl of unlabelled nucleotide mix
   2.5 μl of 10× reaction buffer for RNA polymerase
   2.5 μl of RNase inhibitor (final concentration 1 unit μl^{-1})
   x μl of DNA template (final concentration = 40 μg ml^{-1})
   y μl of appropriate RNA polymerase and *either*
   z μl of ^{3}H-labelled UTP (see note (c) above) *or*
   2 μl of digoxigenin-11-UTP–UTP mix
   w μl of water

   Total volume = 25 μl

2. Incubate for 30 min to 2 h at 37°C.
3. Add 1–2 μl of tRNA (100 mg ml^{-1}; Sigma type XXI), 10 units of RNase-free DNase I and sterile water to a final volume of 100 μl to remove template DNA. Incubate for 10 min at 37°C.
4. For radioactive probes only, remove a 1 μl aliquot to estimate the radioactivity (step 11). Extract the remaining sample once with 100 μl of phenol–chloroform–isoamyl alcohol (25:24:1) and once with chloroform–isoamyl alcohol (24:1).
5. Add an equal volume of 4 M ammonium acetate and 2.5× volume of 100% ethanol and precipitate the RNA on dry ice for 15 min or −20°C overnight.
6. Allow the tubes to warm up to room temperature (to avoid the precipitation of unincorporated nucleotides) then centrifuge for 10 min at 10 000 *g*.
7. Discard the supernatant and wash the pellet by adding 0.5 ml of 70% ethanol while vortexing then spin for 5 min as above (step 6). Remove the supernatant and dry the pellet.
8. Resuspend the pellet in 50 μl of sterile water and add 50 μl of 2× carbonate buffer. Incubate at 60°C for the required time calculated using the following equation:

$$t = \frac{(L_i - L_f)}{K \times L_i \times L_f}$$

where t = time (min), K = rate constant (=0.11 kb min^{-1}), L_i = initial length (kb) and L_f = final length (optimum is 0.15 kb).

9. Add 5 μl of 10% acetic acid and 10 μl of 3 M sodium acetate and 250 μl of 100% ethanol and precipitate, rinse and dry as before (steps 5–7).
10. Resuspend the pellet in 20 μl of sterile water or 1 × TE (Appendix).
11. For radioactively labelled probes remove a 1 μl aliquot and estimate the radioactivity by counting in a scintillation counter. The percentage incorporation can then be calculated by comparing the final counts with the initial counts (step 4).
12. Dilute the probe with 50% formamide in water to 5× the concentration used in the *in situ* experiment (Section 5.4).
13. Store at −80°C.

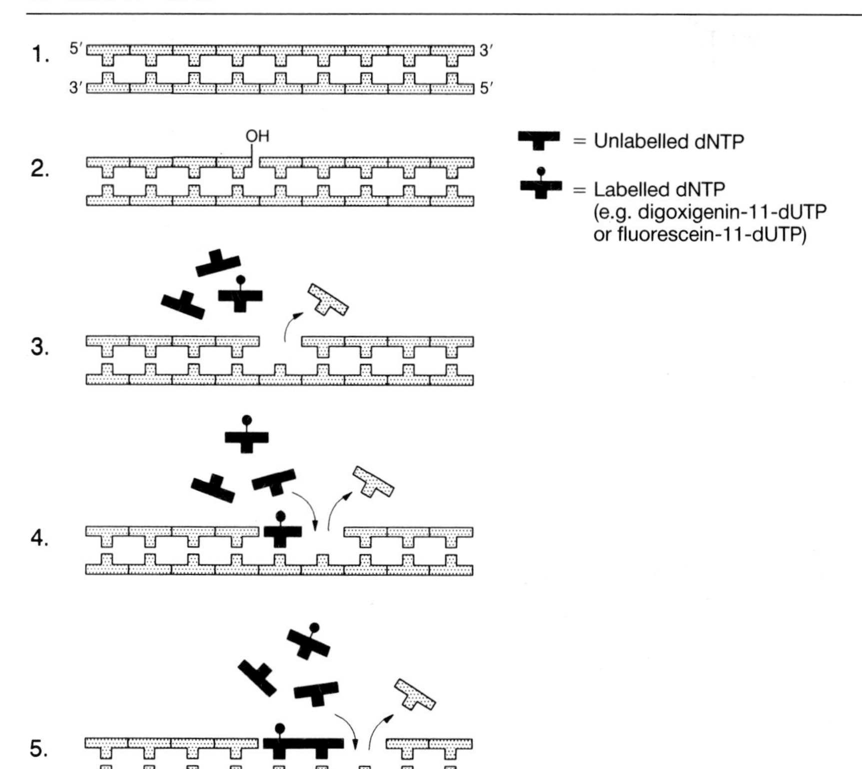

Figure 4.2: The nick translation reaction.

1. The reaction uses two enzymes, DNase I and DNA polymerase I, to incorporate labelled and unlabelled deoxynucleotide triphosphates (dNTPs) into double-stranded DNA.
2. DNase I introduces a single-strand break or 'nick' into the double-stranded DNA to expose a free 3'-OH group.
3. The 5' to 3' exonuclease activity of DNA polymerase I removes mononucleotides at the 5' side of the nick.
4. The DNA polymerase catalyses the incorporation of new dNTPs from solution at the 3'-OH end of the nick.
5. The combined exonuclease and polymerase activities of DNA polymerase I result in the synthesis of a new DNA strand in the 5' to 3' direction. By including labelled dNTPs in the dNTP mix, the newly synthesized complementary strands become labelled.

Table 4.5: Labelling of DNA with biotin-, digoxigenin- or fluorochrome-labelled nucleotides by nick translation

Reagents

(a) 10 × nick translation buffer:
 0.5 M Tris·HCl, pH 7.8 (Appendix)
 0.05 M $MgCl_2$
 0.5 mg ml^{-1} bovine serum albumin, nuclease-free.

(b) Unlabelled nucleotide mixture: dCTP, dGTP and dATP individual nucleotides. Make 0.5 mM solution of each nucleotide in 100 mM Tris·HCl, pH 7.5, and prepare a 1:1:1 mixture.

(c) Labelled nucleotide:
 For digoxigenin: Mix digoxigenin-11-dUTP (1 mM stock solution; Boehringer Mannheim) and dTTP (1 mM stock) to a final concentration of 0.35 mM digoxigenin-11-dUTP and 0.65 mM dTTP.
 For biotin: Use 0.4 mM biotin-11-dUTP (e.g. from Sigma made up from powder in 100 mM Tris·HCl, pH 7.5).
 For fluorochrome-labelled nucleotides: Use either fluorescein-11-dUTP or rhodamine-4-dUTP (1 mM stock; Amersham International) and dTTP (1 mM stock), mix 1:1 fluorochrome dUTP:dTTP for use.

(d) DNA polymerase I/DNase I: 0.4 units μl^{-1} (Gibco BRL).

Method

1. In a 1.5 ml microfuge tube place the following:
 5 μl of 10 × nick translation buffer
 5 μl of unlabelled nucleotide mixture and *either*
 1 μl of digoxigenin-11-dUTP–dTTP mixture *or*
 2.5 μl of biotin-11-dUTP *or*
 2 μl of fluorochrome-labelled nucleotide mixture
 1 μl of 100 mM dithiothreitol
 x μl of DNA equivalent to 1 μg
 y μl of water

Total volume = 45 μl

2. Add 5 μl of DNA polymerase I/DNase I solution, mix gently and centrifuge briefly.
3. Incubate at 15°C for 90 min.

Ethanol precipitation

4. Add 5 μl of 0.3 M EDTA, pH 8, to stop the reaction.
5. Add 5 μl of 3 M sodium acetate (or 5 μl of 4 M LiCl) and 150 μl of ice-cold 100% ethanol.
6. Precipitate the DNA in the freezer (−20°C) overnight or on dry ice for 1–2 h.
7. Spin the tubes at −10°C for 30 min, 12 000 *g*.
8. Discard the supernatant and then wash the pellet by adding 0.5 ml of ice-cold 70% ethanol and then spinning for 5 min as above (step 7).
9. Discard the supernatant and leave the pellet until dry.
10. Resuspend the DNA in 1 × TE (Appendix): 10 μl for genomic probes and 10–30 μl for cloned probes.

Figure 4.3: The oligolabelling reaction.

1. The reaction uses the Klenow fragment of *E. coli* DNA polymerase (5' to 3' polymerase activity) and random oligonucleotides to incorporate labelled and unlabelled deoxynucleotide triphosphates (dNTPs) into DNA.
2. Double-stranded DNA is denatured into single strands and random oligonucleotides are annealed to the single-stranded DNA.
3. The 3'-OH termini of the oligonucleotides serve as primers for the 5' to 3' polymerase activity of the Klenow enzyme. By including labelled dNTPs in the dNTP mix the newly synthesized complementary strands become labelled.

1.

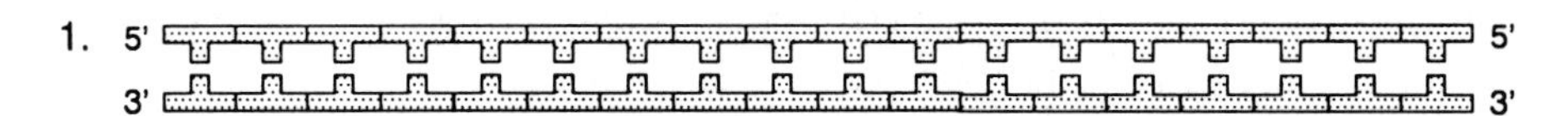

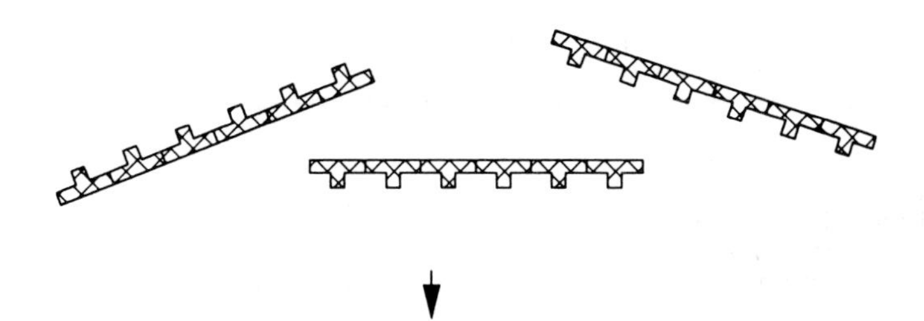

2.

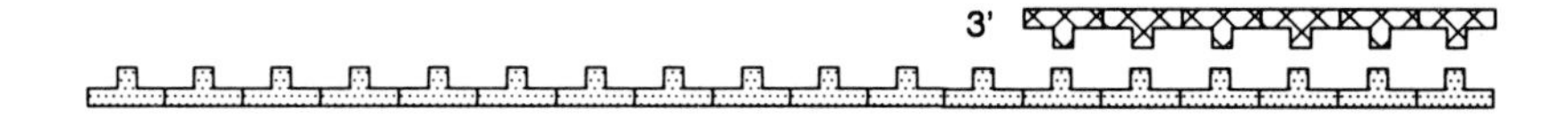

3.

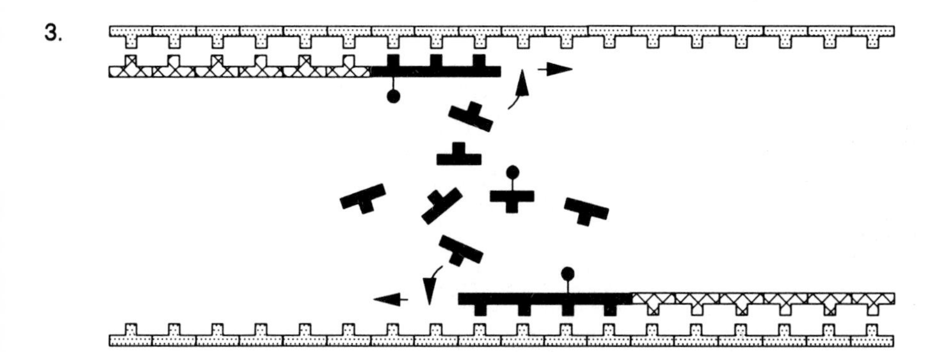

= Random oligonucleotide

= Unlabelled dNTP

= Labelled dNTP (e.g. digoxigenin-11-dUTP or fluorescein-11-dUTP)

1. Less than 200 ng of DNA (and as little as 10 ng) is required for labelling because the DNA is not degraded during the reaction.
2. Either double- or single-stranded DNA can be used as a template for the reaction.
3. Oligolabelling can be used on small fragments of DNA (100–500 bp); nick translation works best on longer fragments of DNA (> 1000 bp).

The main disadvantages of the method are that double-stranded DNA must be denatured, and circular DNA is not efficiently labelled and should be linearized, and the reaction product includes the newly synthesized (labelled) and the template (unlabelled) DNA strands. Unfortunately, the template strand may compete with the labelled strand for probe hybridization sites.

Typically hexanucleotide primers are used, but longer oligonucleotide primers (e.g. 9 bp or 14 bp in length) are now being employed. Longer primers allow the primer concentration to be greatly reduced because they have increased binding efficiency to template DNA. The result is a higher incorporation of label into the DNA. Other developments include the replacement of the *E. coli* Klenow enzyme with bacteriophage T7 DNA polymerase. Experiments have shown that biotinylated nucleotide analogues can be efficiently incorporated into the DNA template using T7 DNA polymerase and random nanomers. Sixty per cent of the biotinylated nucleotides were shown to have been incorporated into the DNA within 10 min.

A protocol for labelling DNA by oligolabelling is given in *Table 4.6*.

Polymerase chain reaction (PCR). A modification of oligolabelling is the PCR. The reaction employs a heat-stable DNA polymerase isolated from the bacterium *Thermus aquaticus* and known as *Taq* polymerase. The enzyme exhibits highly processive 5' to 3' polymerase activity, which is at its maximum at around 72°C. The steps of the reaction are as follows:

1. *Denaturation*: The DNA template is made single-stranded by heating to high temperatures (92–98°C).
2. *Primer annealing*: A pair of specific oligomer primers is annealed to the single-stranded DNA at temperatures between 37°C and 70°C.
3. *DNA synthesis*: The annealed primers serve as a substrate for *Taq* polymerase and DNA synthesis proceeds at between 70°C and 74°C.

These three steps are repeated up to 40 times. Following the first cycle of DNA synthesis the primers can anneal to both the original template DNA and the newly synthesized complementary strand of DNA. By repeating the cycle of denaturation, annealing and synthesis the original target DNA can be amplified many times. Some protocols denature the DNA for 5 min at 91°C and then anneal before adding the enzyme; this ensures that the DNA is fully denatured prior to synthesis without the risk of damaging the enzyme.

Table 4.6: Labelling of DNA with biotin-, digoxigenin- or fluorochrome-labelled nucleotides by oligolabelling

Reagents

(a) 10 × hexanucleotide reaction mixture (in 10 × buffer; Boehringer Mannheim):
0.5 M Tris·HCl, pH 7.2 (Appendix)
0.1 M $MgCl_2$
1 m M dithioerythritol
2 mg ml^{-1} bovine serum albumin, nuclease-free
62.5 A_{260} units ml^{-1} 'random' hexanucleotides.

(b) Unlabelled nucleotide mixture: dCTP, dGTP and dATP individual nucleotides. Make 0.5 mM solution of each nucleotide in 100 mM Tris·HCl, pH 7.5 (Appendix), and prepare a 1:1:1 mixture.

(c) Labelled nucleotide:
For digoxigenin: Mix digoxigenin-11-dUTP (1 mM stock solution; Boehringer Mannheim) and dTTP (1 mM stock) to a final concentration of 0.35 mM digoxigenin-11- dUTP and 0.65 mM dTTP.
For biotin: Use 0.4 mM biotin-11-dUTP (e.g. from Sigma made up from powder in 100 mM Tris·HCl, pH 7.5).
For fluorochrome-labelled nucleotides: Use either fluorescein-11-dUTP or rhodamine-4-dUTP (1 mM stock; Amersham International) and dTTP (1 mM stock). Mix 1:1 fluorochrome dUTP:dTTP for use.

(d) Klenow enzyme: 6 units μl^{-1} (Boehringer Mannheim).

Method

1. Denature linearized DNA (50–200 ng) in boiling water for 5 min and then chill on ice for 5 min.
2. In a 1.5 ml microfuge tube place the following:
3 µl of unlabelled nucleotide mixture
1.5 µl of labelled nucleotide
2 µl of 10 × hexanucleotide reaction mixture
x µl of denatured DNA
y µl of water

Total volume = 19 µl

3. Add 1 µl of Klenow enzyme, mix gently and centrifuge briefly.
4. Incubate at 37°C for 6–8 h or overnight.

Ethanol precipitation

5. Add 2 µl of 0.3 M EDTA, pH 8, to stop the reaction.
6. Add 2 µl of 3 M sodium acetate (or 2 µl of 4 M LiCl) and 60 µl of 100% ethanol.
7. Precipitate the DNA in the freezer (−20°C) overnight or on dry ice for 1–2 h.
8. Spin the tubes at −10°C for 30 min, 12 000 *g*.
9. Discard the supernatant and wash the pellet by adding 0.5 ml of ice-cold 70% ethanol and then spinning for 5 min as above (step 8).
10. Discard the supernatant and leave the pellet until dry.
11. Resuspend the DNA in 1 × TE (Appendix): 10 µl for genomic probes and 10–30 µl for cloned probes.

Taq polymerase has been shown to accept modified nucleotides (e.g. radioactive nucleotides, digoxigenin- or biotin-labelled nucleotides) as substrates. Therefore, the PCR can be used not only to amplify DNA but also produce large quantities of labelled probe DNA that is suitable for *in situ* hybridization. The PCR may become the standard method to label cloned DNA sequences of less than 4 kbp.

A protocol for labelling DNA by the PCR using oligoxigenin is given in *Table 4.7*.

Table 4.7: Labelling of DNA with biotin-, digoxigenin- or fluorochrome-labelled nucleotides by the polymerase chain reaction for DNA sequences inserted into pUC, pUB or other M13-related vectors

Reagents

(a) 10 × PCR buffer:
  100 mM Tris·HCl, pH 8.3 (Appendix)
  500 mM KCl
  30 mM $MgCl_2$
  0.1% gelatin.

(b) Unlabelled nucleotides:
  2.5 mM dATP, dCTP, dTTP, dGTP solution of each nucleotide in 100 mM Tris·HCl, pH 7.5 (Appendix)

(c) Labelled nucleotides:
  Either 1 mM digoxigenin-11-dUTP (Boehringer Mannheim);
  or 0.4 mM biotin-11-dUTP (e.g. from Sigma, made up from powder in 100 mM Tris·HCl, pH 7.5;
  or 1 mM fluorescein-11-dUTP (Amersham International);
  or 1 mM rhodamine-4-dUTP (Amersham International).

(d) M13 primers:
  (1) M13 reverse sequencing primer (17 bases) from Pharmacia, μM solution supplied.
  (2) M13 single-strand primer (17 bases; Pharmacia, 0.2 μM solution supplied).

(e) *Taq* DNA polymerase: 5 units μl^{-1} (Boehringer Mannheim).

(f) DNA: Miniprep DNA should be diluted 1:100 in water; use 3 μl per reaction.

Method

1. Mix in a 1.5 ml microfuge tube:
   5 μl of 10 × PCR buffer
   2 μl of each of dATP, dCTP, dGTP,
   3.25 μl of dTTP
   1.75 μl of labelled nucleotide
   9 μl of M13 single-strand primer
   2 μl of M13 reverse primer
   3 μl of DNA
   23.5 μl of water

   Total volume = 49.5 μl

2. Put grease around the bottom of the tubes and place them in a temperature cycling machine. Run programme 1.

Step	Temp. (°C)	Time (min)
1. (Denaturation)	91	5
2. (Annealing)	47	5

3. Remove each tube individually and add 0.5 μl of *Taq* DNA polymerase, mix and return to the machine.
4. When enzyme has been added to all the tubes, overlay each tube with 50 μl of mineral oil and start programme 2.

Step	Temp. (°C)	Time (min)
1. (Synthesis)	72	*
2. (Denaturation)	91	1
3. (Annealing)	47	1

* Use 1 min for sequences up to 1 kbp, and add 1 min for each additional kbp.

   Repeat cycle 40 times

5. Remove the PCR product from under the mineral oil and ethanol precipitate (see *Table 4.5*, steps 4–9).
6. Resuspend the pellet in 20–30 μl of 1 × TE (Appendix).
7. Check the PCR product by agarose gel electrophoresis. DNA incorporating labelled nucleotides will migrate slower than unlabelled DNA.

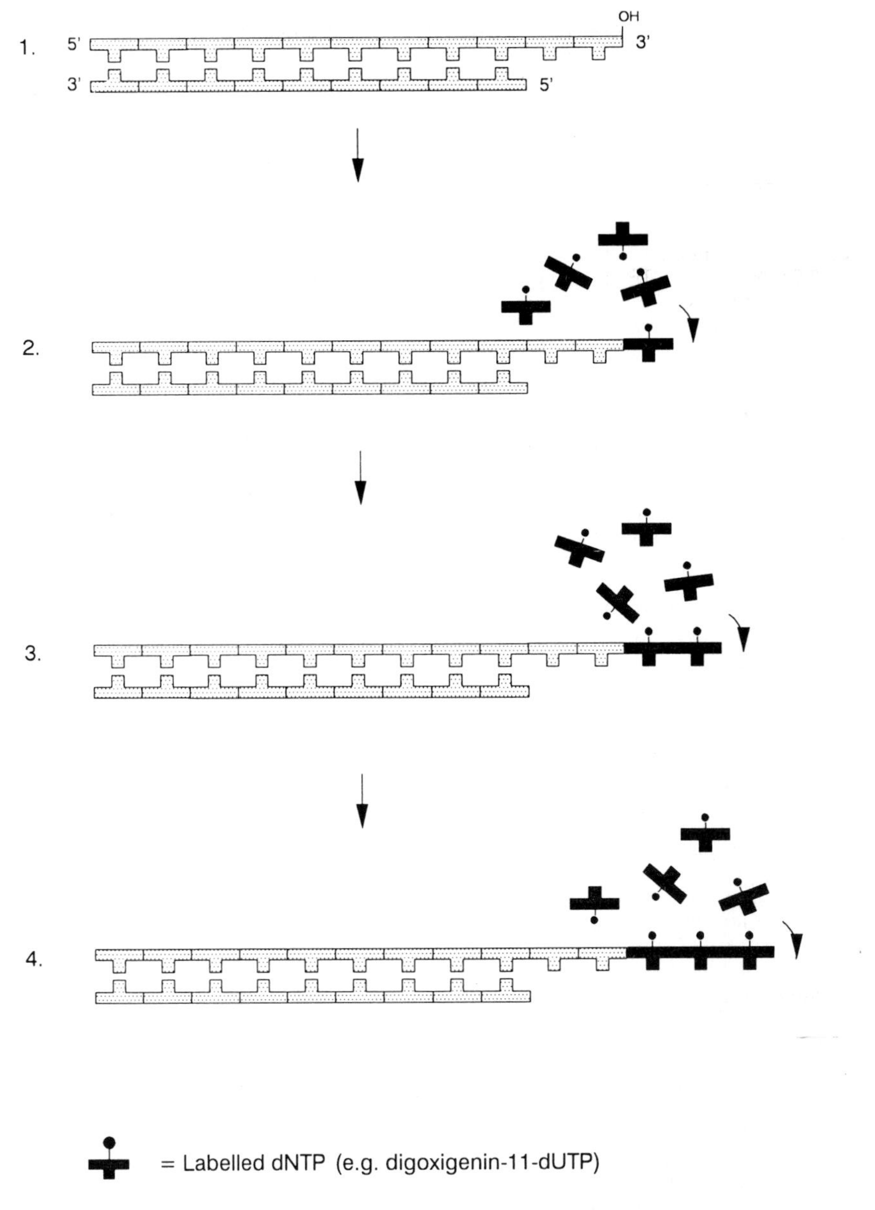

Figure 4.4: The end-labelling reaction.

1. The reaction uses the enzyme terminal deoxynucleotidyl transferase (TdT). The enzyme prefers to use DNA with protruding 3'-termini as acceptors, although blunt or recessed 3'-termini can also be used in the correct conditions.

2–4. TdT catalyses the addition of labelled deoxynucleotide triphosphates (dNTPs) onto the 3'-OH termini of double-stranded (or single-stranded) DNA molecules. The reaction is template independent and the reaction mixture usually contains only the modified dNTP so that the 'tail' produced becomes completely labelled.

End-labelling. The enzyme terminal deoxynucleotidyl transferase (TdT) is an unusual DNA polymerase that will catalyse the addition of nucleotides on to the 3'-OH termini of double- or single-stranded DNA molecules in a template-independent reaction (*Figure 4.4*). The enzyme strongly prefers to use DNA with protruding 3'-termini as acceptors. Blunt or recessed 3'-termini can also be used, although less efficiently, provided that buffers of low ionic strength containing Co^{2+}, Mg^{2+} or Mn^{2+} are used. The extent of labelling depends on the number of 3'-OH groups initially present, since these serve as the initiation sites for the nucleotide addition. Large DNA fragments can be cleaved (e.g. with restriction enzymes or DNase) to increase the number of 3'-OH termini that are available for labelling.

The enzyme will accept modified nucleotides (radioactive nucleotides, biotin-, digoxigenin-labelled nucleotides), and if the reaction mixture contains only the modified dNTP, a 'tail' can be produced that is completely labelled.

The enzyme is particularly useful for labelling oligonucleotide probes (i.e. probes less than 100 bp) for which nick translation and oligolabelling methods are not suitable. The speed of the end-labelling reaction is fast, and the kits that are available (e.g. Boehringer Mannheim) have optimized conditions to enable the length of the 'tail' to be easily controlled.

A protocol for end-labelling DNA with protruding 3'-termini is given in *Table 4.8*.

Table 4.8: End-labelling of oligonucleotide DNA (15–50 bp long) with digoxigenin-, biotin- or fluorescein- labelled nucleotides

Reagents

(a) 5 × DNA tailing buffer (Boehringer Mannheim):
  1 M potassium cacodylate, pH 7.2
  0.125 M Tris·HCl, pH 6.6
  1.25 mg ml^{-1} bovine serum albumin.

Caution: Potassium cacodylate is toxic and should be handled with care.

(b) Cobalt chloride: 25 mM.
(c) dATP: 10 mM dATP in 100 mM Tris·HCl, pH 7.5 (Appendix).
(d) Labelled nucleotide:
  Either 1 mM digoxigenin-11-dUTP (Boehringer Mannheim);
  or 1 mM biotin-11-dUTP (e.g. from Sigma, made up from powder in 100 mM Tris·HCl, pH 7.5);
  or 1 mM fluorescein-11-dUTP (Amersham International).
(e) Terminal deoxynucleotidyl transferase (TdT): 10–15 units μl^{-1} (Boehringer Mannheim).

Method

1. In a 1.5 ml microfuge tube place the following:
   4 μl of 5× DNA tailing buffer
   4 μl of cobalt chloride
   1 μl of labelled nucleotide
   1 μl of dATP
   x μl of DNA equivalent to 250–400 ng
   y μl of water
   Total volume = 19 μl
2. Add 1 μl of TdT, mix gently and centrifuge briefly.
3. Incubate for 15 min at 37°C or 2–3 h at room temperature.
4. Ethanol precipitate as described in *Table 4.6*, steps 5–10.
5. Resuspend the DNA in 10–20 μl 1× TE (Appendix).

4.2.3 Checking incorporation of labels

After labelling it is advisable to check the incorporation of the labelled nucleotides. For radionucleotides this can be measured using a scintillation counter (e.g. *Table 4.4*). Biotin or digoxigenin label incorporation can be measured using a dot blot assay, as described in *Table 4.9*, while the incorporation of fluorochrome-labelled nucleotides can be measured by placing a small amount of probe (e.g. 1 μl) on a glass slide and examining for fluorescence in an epifluorescence microscope with suitable filters (*Table 7.1*).

Table 4.9: Checking for the incorporation of biotin or digoxigenin into the probe using a dot blot

Reagents

(a) Buffer 1:
0.1 M Tris·HCl, pH 7.5
0.15 M NaCl.
(b) Buffer 2: 0.5% (w/v) blocking reagent (Boehringer Mannheim or Amersham International) in buffer 1. This is dissolved by heating the solution to 50–70°C for at least 1 h. The prepared solution can then be stored at 4°C for up to 1 month.
(c) Buffer 3:
0.1 M Tris·HCl, pH 9.5 (Appendix)
0.1 M NaCl
0.05 M $MgCl_2$.
(d) Antibody for the detection of biotin: 1:500 dilution of anti-biotin conjugated to alkaline phosphatase (Vector Laboratories) in buffer 1.
(e) Antibody for the detection of digoxigenin: 1:5000 dilution of anti-digoxigenin conjugated to alkaline phosphatase (Boehringer Mannheim) in buffer 1.
(f) Detection reagents (mix the following immediately prior to use):
22.5 μl of NBT solution (4-nitroblue tetrazolium chloride, 75 mg ml^{-1} in 70% dimethylformamide; available in solution from Gibco BRL). Store frozen in aliquots.
17.5 μl of BCIP (5-bromo-4-chloro-2-indolyl-phosphate, 50 mg ml^{-1} in 70% dimethylformamide available in solution from Gibco BRL). Store frozen in aliquots.
4.96 ml of buffer 3.

Method

1. Cut the Hybond N^+ membrane (Amersham International) to the size required.
2. Soak the membrane in buffer 1 for 5 min and then blot dry between filter paper.
3. Load the DNA onto the membrane (0.5–1 μl) and leave to dry for 5–10 min.
4. Place the membrane in buffer 1 for 1 min and then into buffer 2 for 30 min; shake gently during this period.
5. Drain the membrane slightly, distribute the appropriate antibody over the membrane, and incubate at 37°C for 30 min, shaking gently.
7. Wash the membrane in buffer 1 for 3 × 5 min.
8. Transfer the membrane to buffer 3 for 2 min.
9. Prepare the detection reagents, pour over membrane and leave for 5–10 min in the dark for the colour to develop fully. Wash the membrane in water and air dry.

4.2.4 Chemical labelling

2-Acetylaminofluorene. Both double-stranded and single-stranded DNA and RNA can be chemically labelled with 2-acetylaminofluorene (AAF), which is highly immunogenic (Landegent *et al.*, 1984). AAF is introduced into the nucleic acid by reacting with *N*-acetoxy-2-acetylaminofluorene (*N*-Aco-AAF). The main site of AAF introduction is the C-8 position of the guanine residues. The degree of modification can be controlled by varying the ratio of *N*-Aco-AAF to nucleic acid, however in general, a degree of modification of 5–10% is considered sufficient. Lower degrees of modification (e.g. about 2%) result in a weaker signal, whereas with higher degrees of modification steric factors may interfere with antibody binding during the detection steps.

The reaction is simple and rapid (20–30 min), producing highly stable probes, however its use has perhaps been limited by the toxicity of AAF, which is carcinogenic.

Sulphonation. Single-stranded DNA probes can be chemically labelled by the insertion of a sulphone group at C-6 of cytosine residues. The reaction is catalysed by sodium bisulphite. The resulting sulphone group is stabilized by the substitution of the amine group on C-4 of cytosine with methoxyamine to produce a sulphonate derivative of cytosine that is highly immunogenic (Verdlov *et al.*, 1974). Approximately 10–15% of the cytosine residues become sulphonated during this reaction.

The reaction is simple, and can be performed on unpurified DNA, although DNA less than 100 bp is not efficiently labelled. The reaction is usually conducted overnight but can be performed in 4–6 h by increasing the temperature of the reaction to 42°C and the amount of methoxyamine.

A kit is now available from FMC Bioproducts (Chemiprobe kit) for the labelling and subsequent detection of DNA in this way.

Mercuration. Mercury can be incorporated at the C-5 position of pyrimidine bases (RNA: cytosine, uracil; DNA: cytosine) in nucleic acids (Hopman *et al.*, 1987). By varying the incubation time (usually 8–16 h) the degree of mercury modification can be manipulated. Within 8 h, 30–40% of the uracil and cytosine residues are modified in RNA probes and 40–50% of the cytosine bases in DNA. The toxicity of the procedure has limited its use.

Photolabelling. A number of light-sensitive compounds are now available that can be used to label DNA and RNA; these include photobiotin and photodigoxigenin. These labelling methods have, however, made little impact on *in situ* hybridization although there is no theoretical reason why they cannot be used.

4.2.5 Primed *in situ* labelling

As an alternative to the labelling of nucleic acid followed by probe hybridization *in situ*, a method has been developed called primed *in situ* labelling (PRINS). PRINS involves first annealing a specific DNA sequence to the material; the hybridized DNA then serves as a primer for the incorporation *in situ* of labelled nucleotides by either DNA polymerase or reverse transcriptase. Cloned DNA, synthetic oligonucleotides and PCR products have all been shown to serve as primers for the *in situ* labelling reaction. Direct labelling of chromosomes after primer hybridization has been used to label chromosomes (Koch *et al.*, 1991) and to localize mRNA (Tecott *et al.*, 1988).

Further reading

Baldini A, Ross M, Nizetic D, Vatchvea R, Lindsay EA, Lehrach H, Siniscalco M. (1992) Chromosomal assignment of human YAC clones by fluorescent *in situ* hybridization: use of single-yeast-colony PCR and multiple labelling. *Genomics* **14**, 181–184.

Bauman JGJ, Wiegant J, Borst P, van Duijn P. (1980) A new method for fluorescence microscopical localization of specific DNA sequences by *in situ* hybridization of fluorochrome-labelled RNA. *Exp. Cell Res.* **128**, 485–490.

Cook AF, Vuocolo E, Brakel CL. (1988) Synthesis and hybridization of a series of biotinylated oligonucleotides. *Nucleic Acids Res.* **16**, 4077–4095.

Hopman AHN, Wiegant J, van Duijn P. (1987) Mercurated nucleic acid probes, a new principle for non-radioactive *in situ* hybridization. *Exp. Cell Res.* **169**, 357–368.

Keller GH, Manak MM. (1989) *DNA Probes*. Macmillan Publishers, New York.

Koch JE, Hindkjaer J, Mogensen J, Kϕlvra S, Bolund L. (1991) An improved method for chromosome-specific labeling of α satellite DNA *in situ* by using denatured double-stranded DNA probes as primers in a primed *in situ* labeling (PRINS) procedure. *Genet. Anal. Techniques Applications* **8**, 171–178.

Landegent JE, Jansen in der Wal N, Baan RA, Hoeijmakers JHJ, van der Ploeg M. (1984) 2-Acetylaminofluorene-modified probes for the indirect hybridocytochemical detection of specific nucleic acid sequences. *Exp. Cell Res.* **153**, 61–72.

Lengauer C, Riethman H, Cremer T. (1990) Painting of human chromosomes with probes generated from hybrid cell lines by PCR with *Alu* and L1 primers. *Hum. Genet.* **86**, 1–6.

Lichter P, Cremer T, Borden J, Manuelidis L, Ward DC. (1988) Delineation of individual human chromosomes in metaphase and interphase cells by *in situ* suppression hybridization using recombinant DNA libraries. *Hum. Genet.* **80**, 224–234.

Lichter P, Chang Tang C-J, Call K, Hermanson G, Evans GA, Housman D, Ward DC. (1990) High resolution mapping of human chromosome 11 by *in situ* hybridization with cosmid clones. *Science* **247**, 64–69.

Nisson PE, Watkins PC, Menninger JC, Ward DC. (1991) Improved suppression hybridization with human DNA (Cot-1 DNA) enriched for repetitive DNA sequences. *Focus* **13**, 42–45.

Ried T, Baldini A, Rand TC, Ward DC. (1992) Simultaneous visualization of seven different DNA probes by *in situ* hybridization using combinatorial fluorescence and digital imaging microscopy. *Proc. Natl. Acad. Sci. USA* **89**, 1388–1392.

Sambrook J, Fritsch EF, Maniatis T. (1989) *Molecular Cloning: a Laboratory Manual.* Cold Spring Harbor Laboratory Press, New York.

Selleri L, Hermanson GG, Eubanks JH, Evans GA. (1991) Chromosomal *in situ* hybridization using yeast artificial chromosomes. *Genet. Anal. Techniq. Applic.* **8**, 59–66.

Tecott LH, Barchas JD, Eberwine JH. (1988) *In situ* transcription: specific synthesis of complementary DNA in fixed tissue sections. *Science* **240**, 1661–1664.

Verdlov ED, Monastyrskaya GS, Guskova LI, Levitan TL, Sheichenko VI, Budowsky EI. (1974) Modification of cytidine residues with a bisulphite-*O*-methylhydroxylamine mixture. *Biochim. Biophys. Acta* **340**, 153–165.

Wiegant J, Ried T, Nederlof PM, van der Ploeg M, Tanke HJ, Raap AK. (1991) *In situ* hybridization with fluoresceinated DNA. *Nucleic Acids Res.* **19**, 3237–3241.

5 Denaturation, Hybridization and Washing

5.1 Theory

The *in situ* hybridization reaction exploits the kinetics of nucleic acid duplex formation. Nucleic acid duplexes form by hydrogen bonding between two complementary nucleic acid strands. Each nucleic acid strand is a linear, unbranched polymer consisting of a sugar–phosphate (phosphodiester) backbone with a nucleotide (purine or pyrimidine) linked to each sugar residue. Hydrogen bonds form between the two strands because of the placement of amino (NH_2) and keto (C=O) groups on the complementary nucleotides. In the DNA double helix, adenine (A, a purine) of one polynucleotide strand is hydrogen bonded to thymine (T, a pyrimidine) on the other, while guanine (G, a purine) is hydrogen bonded to cytosine (C, a pyrimidine). In RNA, uracil (U, a pyrimidine) replaces thymine. Purine–purine bonds are too bulky to fit into the helix while pyrimidine–pyrimidine pairs are too far apart.

The stability of the duplex under defined conditions can be determined by calculating its melting temperature (T_m). The T_m is the temperature at which one-half of the duplex molecules become dissociated or 'melted' into single strands. The more stable the hybrid, the higher the T_m (i.e. the greater the energy needed to denature the strands). For DNA duplexes longer than 250 bp in solution the T_m can be calculated using *Equation 5.1*.

$$T_m = 0.41\ (\%\ GC) + 16.6 \log M - 500/n - 0.61\ (\%\ formamide) + 81.5 \quad [5.1]$$

where T_m = 'melting' temperature (°C), (% GC) = percentage of guanine and cytosine in the probe sequence (if unknown, this is usually taken to be 45% for cereals and 40% for human), M = the concentration of monovalent cations (Na^+) in the hybridization solution (mol l^{-1}), n = probe length in base pairs (e.g. usually 250 – 500 bp after a nick translation reaction), % formamide = concentration of formamide expressed as (v/v) percentage. T_m for RNA:DNA hybrids is 10 – 15°C higher. T_m for RNA:RNA hybrids is 20 – 25°C higher.

The stability of the hybrid nucleic acid is dependent on a number of factors:

1. *The proportion of guanine and cytosine (% GC)*. GC pairs are more stable than AT pairs because they are bonded together by three as opposed to two hydrogen bonds. For this reason, DNA with a high GC content is more stable than DNA which is rich in AT, and more energy is required to separate the two strands. A linear relationship exists between DNA hybrid strength and the ratio of AT:GC.
2. *Length of hybrid nucleic acid*. In general, a long duplex is more stable than a shorter one because it is held together by more hydrogen bonds; it thus requires more energy to denature the two strands. Oligonucleotide probes have fewer hydrogen bonds binding the duplex, and this reduces the duplex stability. For example, the T_m of a 30 bp oligonucleotide will be about 5°C lower than for a sequence of more than 250 bp (Lathe, 1990).
3. *The environment of the nucleic acid hybrids*. Typically the hybridization mix (Section 5.4) and the post-hybridization washing solution (Section 5.5) contain formamide in an ionic salt buffer containing monovalent cations. Formamide is a destabilizing molecule which inhibits the formation of nucleic acid duplexes by disrupting the hydrogen bonds. In contrast, increasing the concentration of monovalent cations (i.e. Na^+) serves to stabilize the duplexes. The temperature of these solutions under defined formamide and Na^+ concentrations determines the stringency of the *in situ* hybridization (Section 5.2).
4. *The nucleic acid hybrid*. RNA:RNA hybrids are more thermally stable than DNA:DNA hybrids. DNA:RNA hybrids have intermediate stability.
5. *The presence of mismatched hybrids*. Nucleotides that are incorrectly matched (e.g. T–T rather than A–T) reduce the stability of the hybrid because hydrogen bonds cannot form between them. The extent of the destabilizing effect depends partly on the length of the probe. The longer the hybrid duplex, the less effect a single mismatched pair will have on the stability of the duplex.

In dealing with biological material, the equation cannot be considered as absolute since nucleic acids *in situ* probably do not behave in the same way as in solution. The T_m is likely to be affected by the conformational state of the nucleic acid in the material. DNA within a chromosome shows hierarchical packing: it is coiled into nucleosomes, which are coiled into solenoids and the whole coiled again to form the chromosome fibre. In addition, the type of tissue preparation is likely to affect the duplex stability. For example, fixatives may cross-link nucleic acid-associated proteins so that denaturation must be performed at temperatures higher than that calculated for the T_m. Optimal experimental conditions should be determined empirically for each material investigated.

5.2 Stringency

The stringency at which an *in situ* hybridization experiment is carried out determines the approximate percentage of nucleotides that are correctly matched in the probe and target duplex and is calculated using *Equation 5.2.*

$$\text{Stringency (\%)} = 100 - M_f\,(t_m - t_a) \qquad [5.2]$$

where M_f = mismatch factor (1 for probes longer than 150 bp, 5 for probes shorter than 20 bp), t_m = calculated melting temperature (°C; T_m, *Equation 5.1*) and t_a = temperature (°C) at which the *in situ* hybridization mixture or washing conditions were used.

Stringency is controlled by the temperature, ionic conditions and concentration of helix-destabilizing molecules (e.g. formamide) in the hybridization mix and post-hybridization washing solutions. Under conditions of high stringency only probes with high similarity to the target sequence will be stable enough to remain hybridized. As stringency is lowered the number of mismatched nucleotides (e.g. T–T rather than A–T) that remain hybridized increases. For probes greater than 150 bp the T_m is reduced by about 1°C for each 1% of mismatched pairs, while for oligomers 20 bp long the reduction in T_m can be as high as 5°C for each 1% of mismatched pairs.

The estimated percentage of mismatched nucleotides should be stated in a publication with *in situ* hybridization data because this gives an indication of signal specificity. In practical terms, a precise knowledge of the probe sequence and length may not be known and the target nucleic acids may be packed by coiling with nucleic acid binding proteins such that the precise stringency cannot be calculated and a stringency range of around 5% is usually stated. Schwarzacher-Robinson *et al.* (1988) carefully regulated stringency in experiments designed to detect the location of buoyant gradient satellite sequences in human chromosomes. At low stringency (60–65% probe and target similarity) the probe hybridized to more sites than when the probe was hybridized at higher stringencies (80–85% probe and target similarity; *Figure 2.1h*). Thus regulating stringency can provide some information on the relationship of the sequences at different chromosomal sites.

5.3 Denaturation

For DNA:DNA *in situ* hybridization both the probe and target sequences need to be denatured to make them single-stranded prior to *in situ* hybridization. Riboprobes are single-stranded, but because they sometimes form intramolecular duplexes within similar regions of their sequence they too are usually denatured. Target RNA is immobilized in the cell cytoplasm or nucleoplasm as a single-stranded molecule and does not need denaturing.

Conditions used to denature RNA or DNA probes are usually about 30°C above the estimated T_m of the duplex (conditions for RNA, Section 8.1.2; for DNA, Section 8.2.2). The degree of denaturation is less critical for the probe than it is for the target DNA where there is a narrow window between adequate denaturation and DNA loss. Our results suggest that denaturation of target DNA differs between species, cell types and fixation. Different nucleic acids may be associated with different proteins and be in different conformation states, which can affect denaturation temperatures (Darzynkiewicz, 1990). Overdenaturation leads to DNA loss and hence must be minimized, particularly in the detection of single-copy sequences.

A major area of variation between laboratories is the denaturation procedure for target DNA. Widely used procedures involve acids, alkali or formamide, in ionic buffers at various temperatures, concentrations and durations. For non-radioactive probes in which the label is attached to the probe via an ester linkage, alkali treatment must not be used since it will remove the label by alkali hydrolysis. A few researchers have denatured target DNA using the enzyme exonuclease I (van Dekken *et al.*, 1988). For exonuclease to work satisfactorily DNA must first be nicked, e.g. with DNase I. All techniques give adequate denaturation to allow probe hybridization, but may account, in part, for the wide variation in the quality of *in situ* hybridization seen between laboratories.

In some laboratories slides are denatured separately from the probe, while in others the target and probe are denatured together, a process called combined denaturation. Combined denaturation is our method of choice for LR White sections and chromosome spreads. This method is moderately safe and does not require the handling of large amounts of hot formamide.

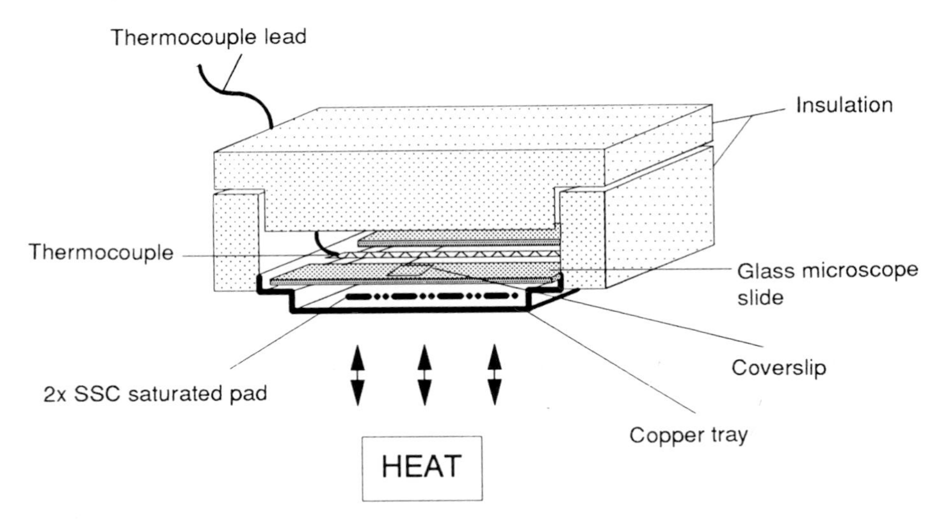

Figure 5.1: Cross-section through the block of the programmable temperature controller modified to hold microscope slides for *in situ* hybridization (Cambio). Reproduced from Heslop-Harrison *et al.* (1991) with permission from Academic Press.

Highly controlled combined denaturation can be achieved using a programmable temperature controller which has been modified to accommodate microscope slides (Heslop-Harrison *et al.*, 1991; *Figure 5.1*). A programmable system enables control of denaturation and hybridization with high accuracy and reproducibility. The schedule outlined in Chapter 8 describes a simpler system that requires only a water bath, but the use of a programmable temperature controller is recommended.

5.4 Hybridization

Once probe and target nucleic acids are single-stranded the probe is typically left to hybridize overnight at 37°C for DNA:DNA hybrids and 50–55°C for RNA:RNA hybrids, which is usually about 20–25°C below the T_m.

For isotopic *in situ* hybridization we use a probe concentration of 0.1–0.3 ng μl^{-1} per kb of the nucleic acid probe. Therefore, the longer the probe the higher the probe concentration. Background radioactive signal can be a problem, and so the final concentration of isotopically labelled probes is more critical than non-radioactively labelled probes. Usually, non-radioactively labelled probes are used at higher concentrations (0.5–2.0 ng μl^{-1} for cloned probes and 1.5–5.0 ng μl^{-1} for genomic probes) and the probe length is not considered in the calculation. However, excessive probe concentrations will cause background signal with all types of probe labels.

Typically, labelled nucleic acid probes are made up in a hybridization mix containing formamide, salts, dextran sulphate, with optional incorporation of blocking DNA or tRNA, sodium dodecyl sulphate and bovine serum albumin.

1. *Formamide* is used in denaturation and hybridization solutions to enable a reaction temperature that is not damaging to tissue morphology. Formamide also regulates stringency.
2. *Salts in solution* are used to regulate the ionic strength of hybridization and denaturation solutions and to help stabilize the nucleic acid duplexes.
3. *Dextran sulphate* is an inert polymer, a polyanion, of high molecular weight (mol. wt = 500 000), which can increase the hybridization reaction rate by a factor of 3. It functions by forming a matrix in the hybridization mixture which concentrates the probe without affecting the stringency. Other polymers (e.g. polyethylene glycol) and non-polymers such as phenol can also increase hybridization rates of DNA.
4. *Unlabelled blocking DNA or tRNA* is included to block probe hybridization to non-specific sites (e.g. cytoplasm). If total genomic DNA is used for blocking it must first be autoclaved into 100–200 bp sized fragments.
5. *Sodium dodecyl sulphate* (SDS) helps in probe penetration by acting as a wetting agent.
6. *Bovine serum albumin* (BSA) can reduce some non-specific probe hybridization.

5.4.1 Hybridization rate

The precise amount of target nucleic acid available for *in situ* hybridization is difficult to assess because of unknown effects of the nucleic acid conformation and its interactions with associated molecules, particularly proteins. However, in most *in situ* hybridization experiments the probe concentration is in excess of the target concentration and the reaction follows first order kinetics.

The rate of hybridization depends on the probe length, complexity of sequence (i.e. number of repeats) and concentration. In general, long probes may result in a slower rate because of limited diffusion into the material. Hybridization rate is increased using dextran sulphate (see above). It is probable that when one strand is immobilized *in situ* the rate of hybridization is reduced by a factor of 7–10 over the time taken for DNA duplexes to form in solution.

5.5 Post-hybridization washing

Post-hybridization washes are usually carried out in a slightly more stringent solution than the hybridization mixture to denature and remove weakly bound probe, leaving only perfectly or nearly perfectly matched nucleotides in the duplex. Typically the washing stringency is about 15–20°C below the T_m of a perfectly matched duplex allowing for about 85% similarity between the probe and target sequences. The stringency required for optimal signal must be determined empirically.

An additional step is often included for RNA:RNA *in situ* hybridization. RNA probes tend to be 'sticky', producing a high level of background signal. This is most effectively removed by a post-hybridization RNase A digestion. RNase A only removes single-stranded, and hence unhybridized, RNA, leaving the nucleic acid duplexes intact.

Further reading

Darzynkiewicz Z. (1990) Acid-induced denaturation of DNA *in situ* as a probe of chromatin structure. *Methods Cell Biol.* **33**, 337–352.

Heslop-Harrison JS, Schwarzacher T, Anamthawat-Jónsson K, Leitch AR, Min S, Leitch IJ. (1991) *In situ* hybridization with automated chromosome denaturation. *Techniques* **3**, 109–115.

Lathe R. (1990) Oligonucleotide probes for *in situ* hybridization. pp. 71–80. In: *In Situ Hybridization, Principles and Practice* (eds JM Polak and JO'D McGee). Oxford University Press, New York.

Meinkoth J, Wahl G. (1984) Hybridization of nucleic acids immobilized on solid supports. *Anal. Biochem.* **138**, 267–284.

Nakamura RM. (1990) Overview and principles of *in situ* hybridization. *Clin. Biochem.* **23**, 255–259.

Raap AK, Marijnen JGJ, Vrolijk J, van der Ploeg M. (1986) Denaturation, renaturation, and loss of DNA during *in situ* hybridization procedures. *Cytometry* **7**, 235–242.

Schwarzacher-Robinson T, Cram LS, Meyne J, Moyzis RK. (1988) Characterization of human heterochromatin by *in situ* hybridization with satellite DNA clones. *Cytogenet. Cell Genet.* **47**, 192–196.

van Dekken H, Pinkel D, Mullikin J, Gray JW. (1988) Enzymatic production of single-stranded DNA as a target for fluorescence *in situ* hybridization. *Chromosoma* **97**, 1–5.

6 Detection of the *In Situ* Hybridization Sites

Following probe hybridization and stringent washing, the sites of probe hybridization are detected. The methods of detection and visualization depend on the type of label incorporated into the probe (*Table 6.1*). Comparisons of the signal-generating systems are shown in *Table 6.2*.

6.1 Detection of radioactively labelled probes

Radioactively labelled probe hybridization sites are usually detected by autoradiography. The slides are first coated with a radiation-sensitive emulsion, which is allowed to dry to give maximum contact between emulsion and specimen. The emulsion is usually 3–4 μm thick, representing a good compromise between sensitivity (increases with film thickness) and resolution (decreases with film thickness). Types and applications of different emulsions are discussed by Baker (1989).

The signal is formed by the interaction of β-particles, emitted from the radioisotope, with atoms in the emulsion. The energy of the interaction reduces silver halides in the emulsion to metallic silver, thereby generating a latent image. The latent image is developed and fixed using standard photographic procedures. The silver grains are visible by light- and dark-field microscopy (Section 7.2.2; *Figure 2.5*, *Figure 2.7d–f*) or electron microscopy (Section 7.3).

The degree of silver grain localization depends on the emission energy of the isotope used (Section 4.2.1, *Table 4.2*). The centre of a cluster of silver grains is assumed to represent a site of probe hybridization. Often in DNA sequence mapping, many chromosome spreads are scored and statistical methods employed to determine which chromosomes show probe hybridization above background levels.

Table 6.1: Detection and signal-generating systems

Label incorporated into DNA or RNA	Detection system	Signal-generating system
1. Radioactive labels		
e.g. ^{3}H, ^{35}S, ^{125}I, ^{32}P	Autoradiography	
2. Non-radioactive labels		
Digoxigenin	Immunocytochemistry	
2-Acetylaminofluorene	Immunocytochemistry	
Sulphone group	Immunocytochemistry	Fluorescence
Mercury/trinitrophenyl ligand	Immunocytochemistry	Enzyme-generated precipitates
Bromodeoxyuridine	Immunocytochemistry	Metals
Biotin	Immunocytochemistry	
	(Strept)avidin	
Fluorochromes	Direct detection	Fluorescence

Table 6.2: Comparisons of signal-generating systems

	Direct label radioisotope		Direct label fluorochrome		Indirect label fluorochrome		Indirect label enzyme precipitate		Indirect label gold conjugate	
LM applicability	***		****		****		****		***	
EM applicability	**			None		None	***	Signal moderate, poor resolution	***	Signal weak, good resolution
Sensitivity	***	1 kb or better	*	>10 kb	**	10 kb	**	10 kb	*	>10 kb
Signal strength	***		**		***		***		**	
Resolution	* LM	Grain scatter	****		***	Loss of resolution on signal amplification	** LM	Some spread of signal	*** LM	
	* EM	Grain scatter					** EM	Some spread of signal	**** EM	
Signal permanence[1]	****		*	Fades on viewing and unstable	*	Fades on viewing and unstable	***		****	
Experimental speed	*** *	Lab time Film exposure	****	Instant results	***		**		***	
Multiple labelling		Impractical	****	Very good colour distinction	***	Good colour distinction, danger of cross-detection	**	Poor colour distinction between precipitates	***	Different grain sizes (EM only)
Counterstains	***	Histochemical stains	****	Chromatin counterstains	****	Chromatin counterstains	***	Histochemical stains	***	Histochemical stains
	**	Chromatin stains					**	Chromatin stains	**	Chromatin stains
Quantitative	***	Count grains	**	Fluorescent signal size and intensity	*	Fluorescent signal size and intensity	*	Colorimetric	***	Count grains (EM only)
Expense	***		**	Epifluorescence microscope	**	Epifluorescence microscope	***		***	
Safety	**	Radiation	***	May be some toxicity	***	May be some toxicity	**	Some carcinogenic compounds	****	

* Poor; ** medium poor; *** medium good; **** good.
[1] Films on EM grids can be very delicate after *in situ* hybridization.

(a)

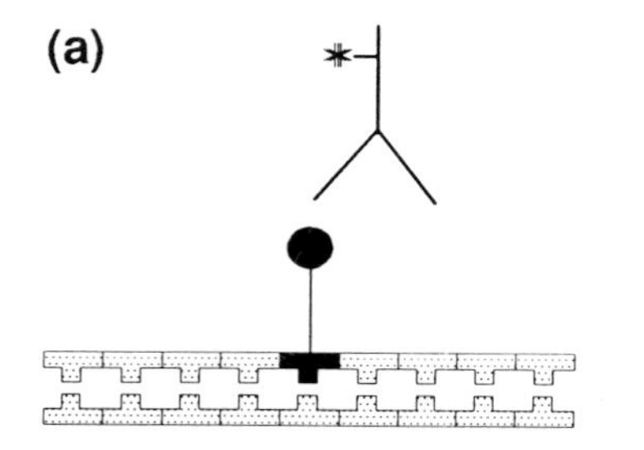

(b)

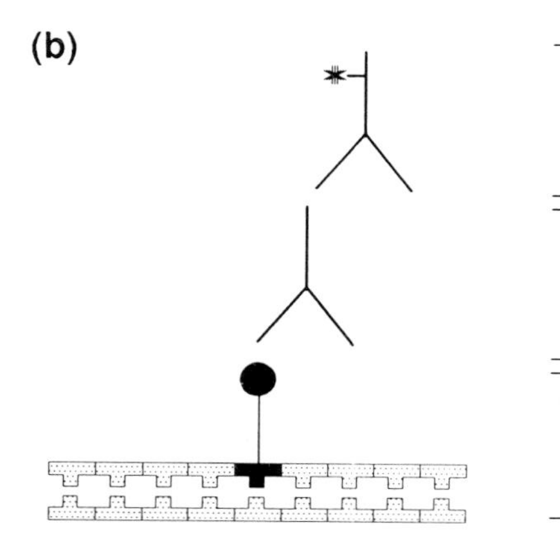

Figure 6.1: Immunocytochemical detection and signal-generating systems for non-radioactive probes. (a) One-step detection system. (b) Two-step detection system.

6.2 Detection of non-radioactively labelled probes

The majority of probe labels are immunogenic and are detected by antibodies raised against the label (e.g. anti-digoxigenin). Biotin has two detection systems, either anti-biotin antibody or the biotin–(strept)avidin system.

6.2.1 Immunocytochemistry

The signal-generating system which enables the sites of probe hybridization to be visualized may be conjugated to the primary antibody (raised against the label) in a one-step detection method (*Figure 6.1a*). Alternatively, a two-step detection method may be used (*Figure 6.1b*) in which a second antibody (raised against the species which produced the primary antibody) carries the signal-generating system. The two-step detection method is generally more sensitive than the one-step approach because several secondary antibodies, each carrying the signal-generating system, can potentially bind to each primary antibody molecule.

For probes labelled with mercury an additional detection step is required since mercury itself is not immunogenic. After hybridization the probe is detected by first reacting the mercury with a ligand containing an immunogenic group such as trinitrophenol (Tnp). The Tnp is detected using anti-Tnp as the primary antibody. A second antibody containing the signal-generating system allows detection of the probe hybridization sites (Hopman *et al.*, 1987). One advantage of this system is that the presence of mercury in the probe does not interfere with probe hybridization and a high degree of mercury incorporation is possible with only limited risk of steric hindrance. Probes labelled with mercury are potentially more sensitive than other non-radioactively labelled probes.

6.2.2 Biotin–(strept)avidin system

Avidin is a glycoprotein extracted from egg white which has a high affinity for biotin. The association constant (K_a = 10^{15} M^{-1}) is about 10^6 times greater than that for antibody–antigen association constants.

The first step in the detection of biotin is the addition of avidin conjugated to a signal-generating system (*Figure 6.2a*). The signal may then be amplified by using biotinylated anti-avidin (an antibody raised against avidin, conjugated to biotin; *Figure 6.2b*) followed by a further layer of avidin conjugated to the signal-generating system (*Figure 6.2c*). Since avidin can bind up to four biotin molecules the potential for amplification is high.

An alternative to avidin is streptavidin, derived from the bacterium *Streptococcus avidini*. Streptavidin is uncharged (unlike the animal-derived avidin) and non-specific electrostatic binding may be reduced. In addition, the molecule lacks carbohydrate groups so non-specific binding to lectins is prevented. However, streptavidin's affinity for biotin is several orders of magnitude lower than that of avidin and it is less stable.

6.3 Signal-generating systems

Visualization of the probe hybridization site depends on a signal-generating system which is conjugated to the antibody or (strept)avidin. Signal-generating systems fall into three main groups: (i) fluorochromes, (ii) enzymes and (iii) metals (*Figure 6.3*). Comparisons of these systems are given in *Table 6.2*. The availability of different systems has enabled the simultaneous detection of several differently labelled nucleic acid sequences (Section 6.4). Using light microscopy different probes may be distinguished

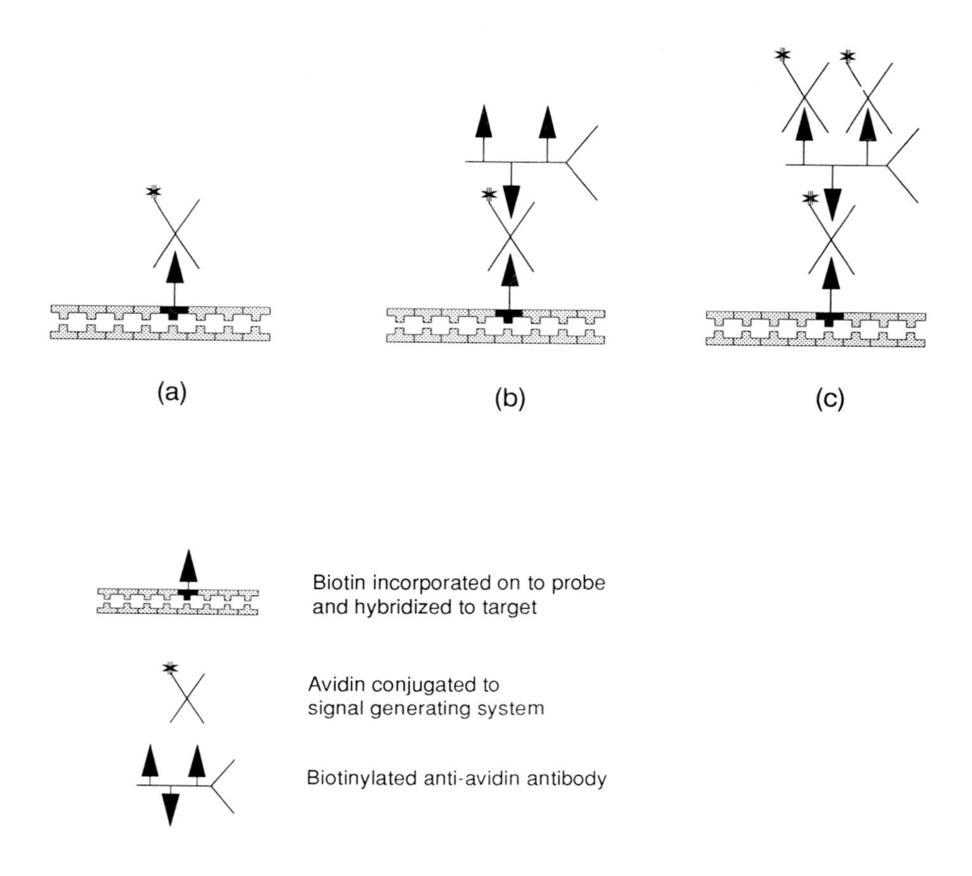

Figure 6.2: Detection of biotin-labelled probes using the biotin–avidin system. (a) Avidin, conjugated to a signal-generating system, binds to the biotin-labelled probe. (b) The signal is amplified by adding biotinylated anti-avidin antibody and then (c) another layer of avidin, conjugated to a signal-generating system, is added.

using fluorochromes with different spectral characteristics (e.g. *Figure 2.1a* and *b*), different enzymes that produce different coloured end products or a combination of both. At the electron microscope level different probes may be detected with different sizes of colloidal gold or by colloidal gold and enzyme-mediated electron-dense precipitates (Section 6.4).

The three main classes of signal-generating systems are described below.

6.3.1 Fluorochromes

Fluorochromes are visualized by excitation with light of the appropriate wavelength (excitation wavelength) and imaging the emitted fluorescence (emission wavelength) using appropriate light filters (Section 7.2.4). A wide variety of fluorochromes are available that can be attached to (strept)avidin or to antibodies. The fluorescent properties of some of the more commonly used fluorochromes (e.g. fluorescein isothiocyanate or FITC, rhodamine and Texas red) are outlined in *Table 6.3*.

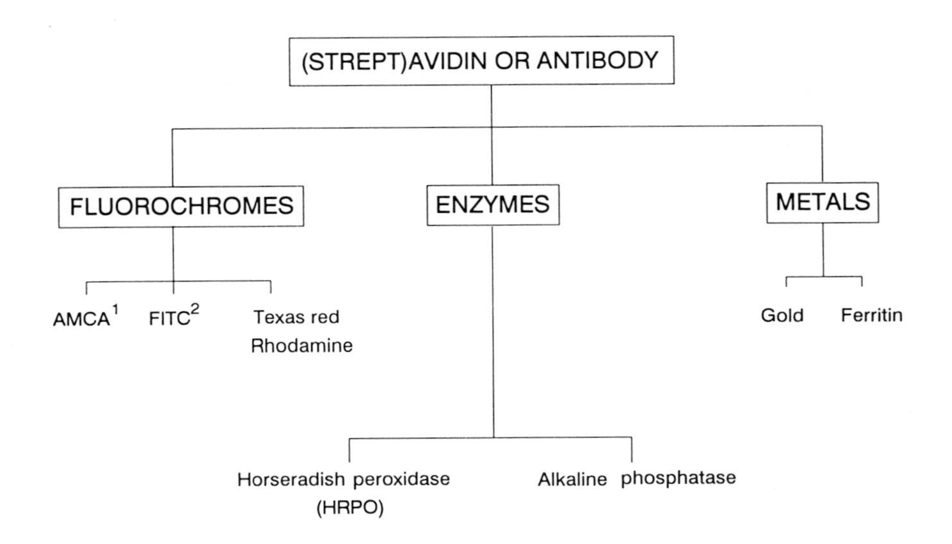

Figure 6.3: Signal-generating systems for non-radioactively labelled probes. 1 = 7-Amino-4-methyl-coumarin-3-acetic acid. 2 = Fluorescein isothiocyanate.

Table 6.3: Fluorochromes

Fluorochrome	Max excitation wavelength (nm)	Max emission wavelength (nm)	Colour of fluorescence
(a) *Signal-generating systems*			
Coumarin AMCA*	350	450	Blue
Fluorescein* FITC	495	515	Green
R-phycoerythrin	525†	575	Red
Rhodamine*	550	575	Red
$Rhodamine_{600}$ TRITC	575	600	Red
Texas red	595	615	Red
$Ultralight_{680}$	red	680	Far red
Cy 5	648	665	Far red
(b) *DNA counterstains*			
Chromomycin A3	430	570	Yellow
DAPI	355	450	Blue
Hoechst 33258	356	465	Blue
Propidium iodide	340,530	615	Red

* Available conjugated directly to nucleotides (Amersham International).
† R-phycoerythrin has a broad excitation range, 450–570 nm.

In some cases fluorescent detection is preferred over enzymatic systems because of better spatial resolution (e.g compare *Figure 2.1c* and *d*), the ability to quantify fluorescent signals by photon counting and the greater potential for simultaneous detection of more than one probe (e.g. *Figure 2.1a* and *b* and Nederlof *et al.* (1990); see Section 6.4). New fluorochromes, based on cyanins that emit in the infrared, can now extend the spectral range still further. Under development are fluorochromes that can be excited at the same wavelength of light but emit light at different wavelengths, allowing for easy detection of different labels at the same time.

6.3.2 Enzyme-mediated reporter systems

Enzyme-mediated reporter systems work by catalysing the precipitation of a visible product at the hybridization site. Many enzymes commonly used in immunocytochemistry are available conjugated to either (strept)avidin or an antibody; these are listed in *Table 6.4*. The most common are horseradish peroxidase and alkaline phosphatase.

The choice of enzyme depends on the material examined. Some tissues contain endogenous enzyme activity; if this is not blocked it can lead to unacceptable background labelling and confusing results. In tissues in which endogenous peroxidase is a problem (e.g. erythrocytes, neutrophils, macrophages) endogenous peroxidase activity may be blocked with periodate and borohydride (Heyderman, 1979), sodium nitroferricyanide (Straus, 1971) or phenyl hydrazine (Straus, 1972). Endogenous alkaline phosphatase is present in placental and intestinal tissue. In fresh placental tissue, endogenous alkaline phosphatase activity may be blocked by incubating the sections in levamisole (Ponder and Wilkinson, 1981). These blocking methods are not suitable for intestinal tissue where endogenous alkaline phosphatase can be removed with 20% acetic acid. When unacceptable levels of endogenous enzyme activity occurs, it is best to change the signal-generating system rather than expend effort removing the endogenous activity.

The major advantages of using enzyme-mediated detection systems are the stability of the signal and the simplicity and cost of the light microscope needed to visualize the signal.

Table 6.4: Enzymes used in signal-generating systems

Enzyme	Substrate	Colour of product
Horseradish peroxidase	Diaminobenzidine (DAB)	Brown
	AEC*	Brick red
Alkaline phosphatase	BCIP/NBT	Blue
	Vector red†	Red

* 3-Amino-9-ethylcarbazol.
† Available from Vector Laboratories.

Horseradish peroxidase. Horseradish peroxidase (HRPO) is usually attached to a secondary antibody (two-step method; *Figure 6.1b*), and the most common substrate for the enzyme is diaminobenzidine (DAB, a carcinogen). The reaction produces a brown deposit that is well localized and exhibits bright reflectance properties (Section 7.2.3). The signal can be enhanced by silver amplification (*Figure 2.2*; Section 8.6; Manuelidis and Ward, 1984). The DAB deposit is electron dense and is thus suitable for detecting probe hybridization sites by electron microscopy (*Figure 2.3*).

The brown precipitate provides a clear contrasting colour to routine nuclear counterstains. Brown also contrasts with the blue precipitate formed using a reaction mediated by alkaline phosphatase, enabling more than one probe to be detected simultaneously. The distinction between the different coloured precipitates is not, however, always as clear as between different fluorochromes (Section 7.2.2).

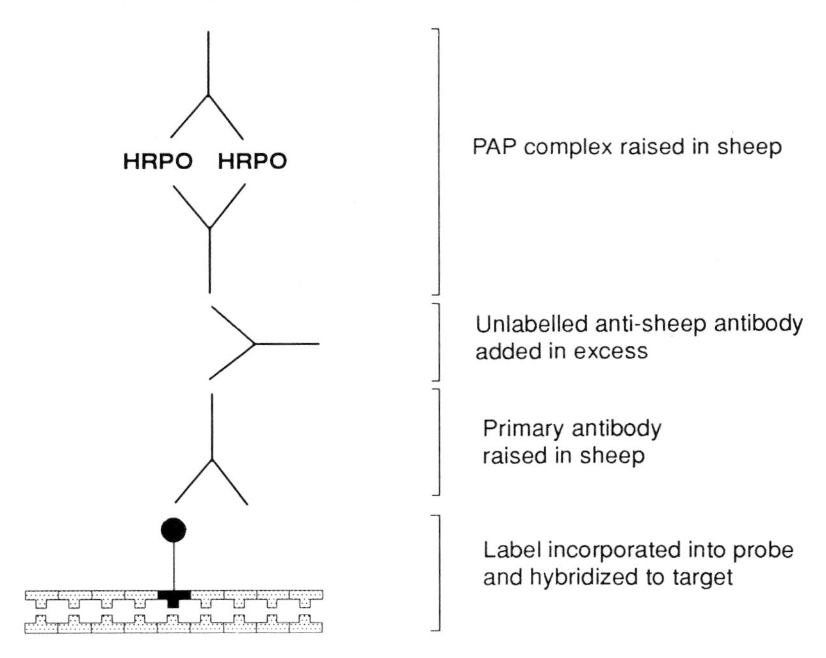

Figure 6.4: Detection of non-radioactively labelled probes using the PAP complex. HRPO, horseradish peroxidase.

Sensitivity of probe detection can be increased by using the peroxidase–anti-peroxidase (PAP) method (*Figure 6.4*). PAP is a preformed antibody–HRPO complex bound immunogenically. The probe hybridization sites are detected by first using an unlabelled primary antibody raised against the probe label. Then a secondary antibody is added in excess so that only one of its antibody binding sites is bound to the primary antibody. The other binding site is free to bind to the antibody component of the PAP complex. The primary antibody and the antibodies in the PAP complex must be raised in the same animal species.

Alkaline phosphatase. The most sensitive substrate for alkaline phosphatase is the BCIP/NBT (5-bromo-4-chloro-3-indolylphosphate/nitroblue tetrazolium) reaction, which generates a blue precipitate at the site of probe hybridization. Kits are now available from Vector Laboratories using alkaline phosphatase to produce more than one coloured precipitate. An alkaline phosphatase–anti-alkaline phosphatase (APAAP) complex that is used in a similar way to the PAP complex (*Figure 6.4*) enables an increase in the sensitivity of detection to be achieved. Alkaline phosphatase reaction products should not be mounted in DPX mountant (BDH) since this can make the coloured precipitate granular and can change its colour. Material should be mounted in Euparol (BDH) after rapid dehydration through an ethanol series (15–30 sec in each solution).

6.3.3 Metal reporter systems for light and electron microscopy

The most widely used metal for *in situ* hybridization is colloidal gold, which is available conjugated to (strept)avidin and antibodies. Colloidal gold can be visualized at both the light and electron (*Figure 2.4*) microscope level. Detection sensitivity using gold probes at the light microscope level can be increased either chemically by silver enhancement (Holgate *et al.*, 1983) or by visualization using reflection contrast microscopy (Section 7.2.3).

At the electron microscope level, colloidal gold offers a number of advantages over enzyme-mediated electron-dense precipitates (e.g. HRPO/DAB described above). Colloidal gold is available in a number of discrete sizes (e.g. 1 nm, 5 nm, 10 nm, 15 nm and 20 nm) and, by using a combination of sizes to detect different probes, more than one sequence can be detected simultaneously (McFadden *et al.*, 1990). In addition, the discrete spherical structure of each particle of colloidal gold enables some degree of signal quantification and the highest resolution of signal detection (Section 7.4). However, the sensitivity of this detection system is usually lower than for other detection methods (e.g. HRPO/DAB), which may be a problem for localizing low-copy or dispersed sequences.

Other metal reporter systems (including ferritin and haemocyanin) are available but are less widely used, possibly because their sizes are less precisely controlled than colloidal gold.

6.4 Multiple sequence detection

Multiple labelling strategies are very important to determine the relationship of sequences to each other, to identify chromosomes simultaneously with a new probe sequence and to localize many probes simultaneously. Multiple labelling experiments employ more than one probe, each of which is differently labelled. The simplest multiple labelling experiments involve incorporating different fluorochrome-conjugated nucleotides (Amersham International) directly into each probe (e.g. rhodamine-conjugated dNTP into one probe and fluorescein-conjugated dNTP into another; *Figure 2.1b*). This experiment is simple because after probe hybridization and washing the different sites of probe hybridization can be identified immediately by their fluorescence colour. Different probes can also be labelled with, for example, digoxigenin, biotin or mercury and appropriate detection systems employed.

To conduct simultaneous *in situ* hybridization of differently labelled probes, each labelled probe is mixed at the correct concentration into the same probe hybridization mix. Denaturation, hybridization and post-hybridization washes are as described (RNA, Sections 8.1.2–8.1.4; DNA, Sections 8.2.2–8.2.4). Multiple labelling using, for example, a biotinylated probe and a digoxigenin-labelled probe requires two different detection systems. The optimum signal-generating systems for multiple labelling are fluorochromes. When choosing the fluorochromes ensure that the fluorescent spectra for excitation and emission do not overlap because too much spectral overlap will blur the distinction between the probe signals (*Table 6.3*).

For the simultaneous detection of a biotin-labelled probe with Texas red (red fluorescence) and a digoxigenin-labelled probe with fluorescein (green fluorescence, as in *Figure 2.1a*) the steps described in Section 8.4 are carried out with the following modifications:

1. Detection. In step (iii) mix Texas red–avidin conjugate (5 $\mu g\ ml^{-1}$) with fluorescein–anti-digoxigenin conjugate (5 $\mu g\ ml^{-1}$) in BSA block.
2. Amplification of signal. In step (vi) mix biotinylated anti-avidin (5 $\mu g\ ml^{-1}$) with FITC-conjugated anti-sheep (25 $\mu g\ ml^{-1}$) in normal goat serum block.
3. For counterstaining (Section 8.8.1), propidium iodide should not be used as the spectral properties of propidium iodide overlap with Texas red.

There is now the option for really elegant multiple labelling strategies by detecting sites of probe hybridization with several different fluorochromes. By carefully choosing fluorochromes with complementary excitation and emission spectra, and using more than one label in some probes, up to 20 different YAC clones have been simultaneously localized on the same human chromosomes to form 'chromosomal bar codes' (Lengauer *et al.*, 1993). Detecting this many sequences requires combinatorial labelling of probes (i.e. incorporating more than one label in each probe in different ratios) and also an epifluorescence microscope equipped with a digital camera and computer software. The software is used to pseudocolour the signal depending on the absolute fluorescence and ratio of fluorescence at different wavelengths.

Multiple labelling strategies can also be conducted on material examined by electron microscopy (McFadden *et al.*, 1990). This is achieved using, for example, 5 nm gold conjugated to streptavidin (for biotin-labelled probes; e.g. Amersham International or Biocell) together with 15 nm gold conjugated to anti-sheep antibodies (for digoxigenin-labelled probes).

As an alternative strategy of detecting different probes to the same material, Heslop-Harrison *et al.* (1992) have shown that material can be reprobed. Sites of probe hybridization detected using fluorochromes can be identified and photographed. The material is then washed, covered with a new probe mixture and the denaturation, hybridization and detection steps repeated. The sites of hybridization of the new probe can be compared with the first probe on photographs or by aligning and displaying the images together using digital imaging (Section 7.2.6).

Further reading

Baker JRJ. (1989) Autoradiography: a comprehensive overview. *RMS Microscopy Handbook* Vol. 18. Oxford University Press, New York.

Heslop-Harrison JS, Harrison GE, Leitch IJ. (1992) Reprobing of DNA: DNA *in situ* hybridization preparations. *Trends Genet.* **8**, 372–373.

Heyderman E. (1979) Immunoperoxidase technique in histopathology: applications, methods and controls. *J. Clin. Pathol.* **32**, 971–978.

Holgate CS, Jackson P, Cowen PN, Bird CC. (1983) Immunogold silver staining: a new method of immunostaining with enhanced sensitivity. *J. Histochem. Cytochem.* **31**, 938–944.

Hopman AHN, Wiegant J, van Duijn P. (1987) Mercurated nucleic acid probes, a new principle for non-radioactive *in situ* hybridization. *Exp. Cell Res.* **169**, 357–368.

Lengauer C, Speicher MR, Popp S, Jauch A, Taniwaki M, Nagaraja R, Riethman HC, Donis-Keller H, D'Urso M, Schlessinger D, Cremer T. (1993) Chromosomal bar codes produced by multicolor fluorescence *in situ* hybridization with multiple YAC clones and whole chromosome painting probes. *Hum. Molec. Genet.* **2**, 505–512.

McFadden G, Bönig I, Clarke A. (1990) Double label *in situ* hybridization for electron microscopy. *Trans. Roy. Microscop. Soc.* **1**, 683–688.

McNeil JA, Johnson CV, Carter KC, Singer RH, Lawrence JB. (1991) Localizing DNA and RNA within nuclei and chromosomes by fluorescence *in situ* hybridization. *Genet. Anal. Techniq. Applic.* **8**, 41–58.

Manuelidis L, Ward DC. (1984) Chromosomal and nuclear distribution of the HindIII 1.9 kb human DNA repeat segment. *Chromosoma* **91**, 28–38.

Nederlof PM, van der Flier S, Wiegant J, Raap AK, Tanke HJ, Ploem HJ, van der Ploeg M. (1990) Multiple fluorescence *in situ* hybridization. *Cytometry* **11**, 126–131.

Polak JM, van Noorden S. (1987) An introduction to immunocytochemistry: current techniques and problems. *RMS Microscopy Handbook* Vol. 11. Oxford University Press, New York.

Ponder BA, Wilkinson MM. (1981) Inhibition of endogenous tissue alkaline phosphatase with the use of alkaline phosphatase conjugates in immunohistochemistry. *J. Histochem. Cytochem.* **29**, 981–984.

Singer RH, Lawrence JB, Villnave C. (1986) Optimization of *in situ* hybridization using isotopic and non-isotopic detection methods. *Bio-Techniques* **4**, 230–250.

Straus W. (1971) Inhibition of peroxidase by methanol and by methanol–nitroferricyanide for use in immunoperoxidase procedures. *J. Histochem. Cytochem.* **19**, 682–688.

Straus W. (1972) Phenylhydrazine as inhibitor of horseradish peroxidase for use in immunoperoxidase procedures. *J. Histochem. Cytochem.* **20**, 949–951.

Trask BJ. (1991) Fluorescence *in situ* hybridization: applications in cytogenetics and gene mapping. *Trends Genet.* **7**, 149–154.

7 Imaging Systems and Analysis of Signal

7.1 Aims

The visualization method aims to localize the *in situ* hybridization signal with the maximum sensitivity and spatial resolution possible.

Many of the limitations of *in situ* hybridization can be tackled not only by improvements in experimental protocols but also by careful choice of visualization methods. For example, low signal strength can be compensated for by sensitive visualization techniques (e.g. low light-sensitive cameras; Section 7.2.6) as well as by experimental amplification of signal. Low contrast can be improved by changes in the microscopy technique and careful use of filters. Other problems can be tackled by data analysis. For example, high backgrounds or low signal resolution can be overcome by statistical data analysis.

Consideration of the experiment before it starts will determine whether the final analysis will be in the light or electron microscope. The latter will be preferred where considerable sub-light microscopic detail (e.g. interphase strands of chromatin or nucleolar substructure) or higher resolution of signal location is required. For DNA:DNA *in situ* hybridization to chromosomes and nuclei and for subcellular localization of mRNA much work is near the limits of resolution of the light microscope. Increasingly, workers are using electron microscopy to improve the resolution of the signal.

For light microscopy the method chosen depends on a wide range of parameters, although the type of signal-generating system used for detecting the *in situ* signal is normally the key factor. Thus, if a fluorescent detection system is used, only epifluorescence light microscopy is suitable.

7.2 Light microscopy

7.2.1 Before *in situ* hybridization

Light microscopy is important for prescreening of slides before the hybridization steps to ensure that the quality of the material is high.

Chromosome preparations. Any chromosome preparation needs to be checked before the *in situ* procedure to ensure that the quality of the preparation is suitably high; a poor slide will give poor results! The examination can be carried out with unstained chromosomes by phase-contrast microscopy on dry or xylene-mounted chromosomes. It is often worth recording coordinates of good metaphases to check that loss, particularly selective loss, of good, isolated (without underlying cytoplasm) metaphase chromosomes, does not occur during the *in situ* hybridization procedure.

Where nuclei are very small (e.g. the plant *Arabidopsis* or the animal *Caenorhabditis*) the preparations may be scored before the *in situ* hybridization procedure by staining with the DNA-binding fluorochrome 4′, 6-diamidino-2-phenylindole (DAPI) (Section 8.8.1). DAPI is removed during post-fixation and the *in situ* procedure without significantly interfering with probe hybridization.

Tissue sections. To check that the material has been adequately fixed and embedded, tissue sections should be stained and examined prior to *in situ* hybridization using transmitted light microscopy (Section 8.7.2). Sections for RNA *in situ* hybridization can be stained with acridine orange to test for RNA retention (staining conditions for acridine orange are described by Darzynkiewicz, 1990).

7.2.2 Imaging *in situ* hybridization signal using transmitted light microscopy

Radioactive label detection. Transmitted light microscopy is the preferred method for visualizing *in situ* hybridization sites following autoradiography (Section 6.1). The material can be visualized by staining (including cytological banding methods for chromosomes) or by phase contrast. Staining before adding the emulsion is not recommended as some stains, under certain conditions, can emit photons. In addition, some stains are eluted by the developing and fixing procedures of the emulsion.

Filters are normally used to increase contrast between silver grains and the background. A red filter is preferred for examining silver grains in the emulsion, while filters complementary in colour to the stain are used to image the material (e.g. a green filter for red counterstains).

Dark-ground microscopy, in which silver grains appear as bright dots on a black background, can be useful for examining large clouds of silver grains and to determine the levels of background labelling (*Figure 2.7d* and *f*).

Detection of* in situ *hybridization signal generated by enzyme-mediated reporter systems. For visualizing the coloured products precipitated by enzyme-mediated reporter systems (Section 6.3.2), transmitted light microscopy is usually used. As with radioactive label detection, contrast of the signal can be enhanced by using filters complementary in colour to the coloured precipitate and material counterstain. Many of the coloured precipitates used to detect *in situ* hybridization sites absorb light of many wavelengths and do not have a clear, easy to determine colour like fluorochromes. For example, precipitates from diaminobenzidine (DAB) can vary from a deep brown–black colour to a pale straw colour. This can make the simultaneous detection of more than one enzyme-mediated precipitate difficult.

All counterstains (e.g. Giemsa for chromosome analysis or eosin and haematoxylin for paraffin wax sections) should be used lightly so as not to obscure *in situ* hybridization signal.

7.2.3 Reflection contrast microscopy

This type of microscopy uses white light to illuminate the slide from above (epi-illumination) and a special reflection contrast objective lens to focus polarized light onto the specimen. Where signal, material and background have different reflectance properties, a high-contrast image can be generated. Biological material appears dull or near black while colloidal gold and the enzyme-mediated precipitate from DAB and BCIP/NBT have particularly high reflectance properties and the sites of *in situ* hybridization can appear bright (*Figure 2.7b*). The method can give better contrast than transmitted light microscopy and can be extremely sensitive, enabling the detection of low levels of *in situ* hybridization signal (Landegent *et al.*, 1985).

7.2.4 Epifluorescence microscopy

The principle behind fluorescence microscopy is that a photon of a particular wavelength (excitation wavelength) excites an electron in the fluorochrome, making it jump into an outer electron shell. The excited electron is unstable and on returning to its ground (stable) state looses energy, which is emitted as light (fluorescence). According to Stoke's law the wavelength of emitted light is always longer than the excitation wavelength (*Table 6.3*).

The light source for epifluorescence microscopy is usually an ultrahigh-pressure mercury vapour lamp (particularly 50 or 100 W) which emits

Table 7.1: Filters used to visualize fluorochromes by epifluorescence microscopy

Fluorochrome	Excitation colour	Excitation filter	Chromatic beam splitter	Barrier filter	Fluorescence colour
DAPI	Ultraviolet	G365 or BP340–380	CBS420	LP420	Blue
FITC	Blue/violet	BP450–490	CBS510	LP520	Green
Texas red	Green	BP536–556 or BP515–560	CBS580	LP590	Red
Propidium iodide	Green	BP536–556 or BP515–560	CBS580	LP590	Red

BP, bandpass filter; wavelengths between the numbers shown are transmitted.
CBS, chromatic beam splitter; wavelengths less than number indicated are reflected while those greater than the number are transmitted.
G, solid glass filter; these work in a similar way to the bandpass filters by transmitting only a band of wavelengths, however they are less efficient than bandpass filters. Wavelengths around the number are transmitted.
LP, longpass filter; wavelengths longer than this number are transmitted.

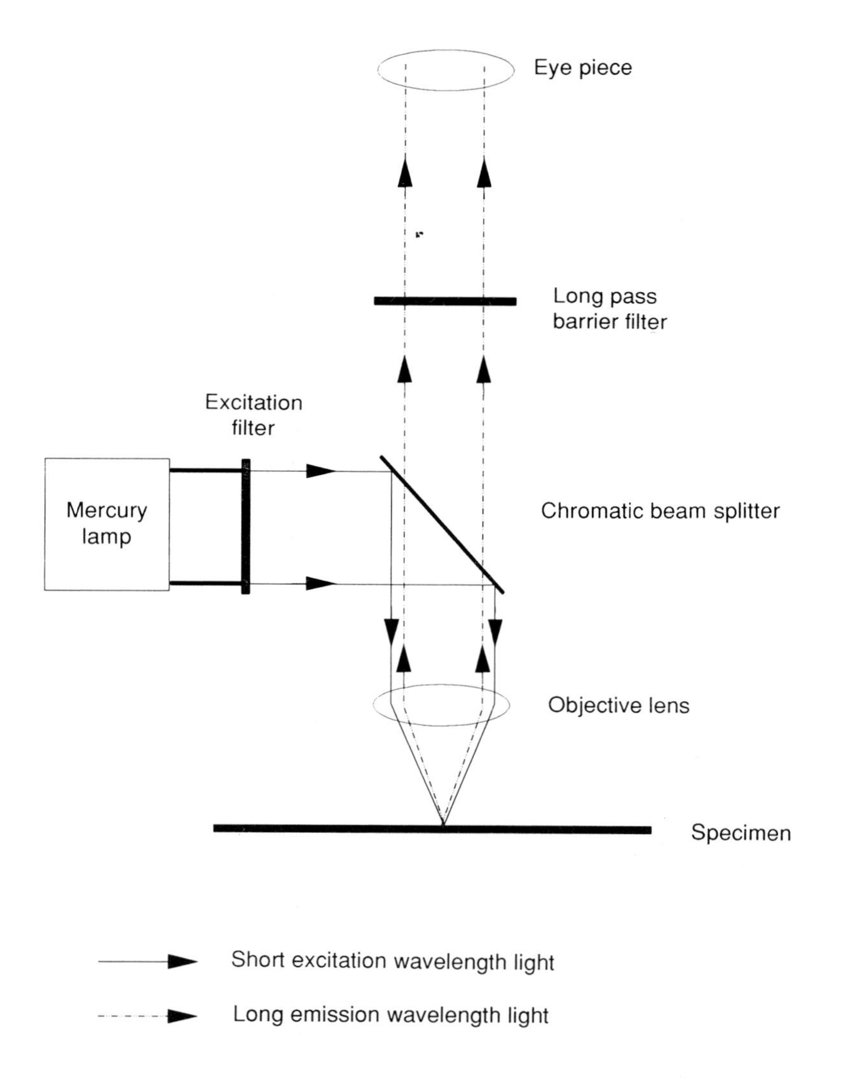

Figure 7.1: The fluorescence microscope.

ultraviolet, visible and infrared light. The microscope (*Figure 7.1*) has excitation filters (*Table 7.1*) to select the correct wavelengths of light for the particular fluorochrome (*Table 6.3*). The selected wavelengths are focused on the specimen using the objective lens. The excitation filter may be a bandpass filter that only transmits light within a narrow, defined range of wavelengths. Alternatively, a short-pass filter may be used, which only transmits light below a certain wavelength. The emitted light is imaged through a long-pass barrier filter, which only transmits light longer than a certain wavelength. In order for the microscope to operate successfully, an additional filter is needed between the excitation filter and the long-pass filter; this is called the chromatic beam splitter. The chromatic beam splitter is positioned between the excitation filter and the specimen at 45° and reflects short, excitation wavelength light on to the specimen. The fluorescent light, emitted from the specimen at a longer wavelength, is nearly fully transmitted through the chromatic beam splitter to the long-pass filter. The chromatic beam splitter is important as it allows the excitation filter and the long-pass filter to lie in different light paths above the specimen.

Fluorochromes are readily bleached by epifluorescence microscopy, although the speed of bleaching can be reduced using anti-fade reagents (Section 8.8.2). Excitation light of high energy (short wavelength) increases the speed of fluorochrome bleaching. It is therefore important when visualizing two fluorochromes to examine the fluorochrome excited at the longer wavelength first (e.g. examine Texas red signal before DAPI signal). The problem of bleaching coupled with the low intensity of emitted light requires the use of high-performance objective lenses that transmit UV light (i.e. a high numerical aperture with good correction for spherical and chromatic aberrations). The recording of fluorescent images is discussed in Section 7.2.6.

Epifluorescence microscopy has progressed enormously with the introduction of high-performance interference filters in which the filter effect is based on light interference rather than absorption. Interference filters consist of vapour-deposited layers (a dielectric or metal) on glass or quartz supports. The transmission and reflectance properties of these filters is determined by the thickness, number, composition and sequence of deposition of the layers. These interference filters can be designed to transmit or block certain wavelengths of light very specifically and effectively.

Another important advance is the development of dual and triple band filter sets, which enable the simultaneous visualization of two or three fluorochromes, respectively (e.g. Johnson *et al.*, 1991). These filter sets are becoming particularly important for the physical mapping of closely spaced probes because the image shift from one single band filter set to another is sufficient to cause loss of precise positional information.

7.2.5 Confocal microscopy

Conventional epifluorescence microscopy can suffer from a distorted image generated from out-of-focus information interfering with the image. This is particularly true when the signal is imaged within intact cells or cell layers. The confocal scanning optical microscope or confocal microscope offers advantages in that it can remove much of the out-of-focus information, enabling non-invasive optical sectioning. A complete series of optical sections through biological material enables a study of the three-dimensional distribution of *in situ* hybridization signal in biological material (e.g. *Figure 3.1*; Rawlins and Shaw, 1990).

The confocal microscope (*Figure 7.2*) scans the specimen with a spot of laser light focused through a conventional epifluorescence microscope. The excitation wavelength of the light excites fluorochromes used in the signal-generating systems (usually rhodamine, Texas red or FITC). The emitted light passes to the confocal aperture that is confocal with the 'in focus' information. Most of the light that passes through the aperture is in focus, while most light from above and below the focal plane is defocused at the aperture and is prevented from reaching the photomultiplier. The photomultiplier, receiving mainly focused light, displays the image point by point on to an appropriate high-resolution monitor.

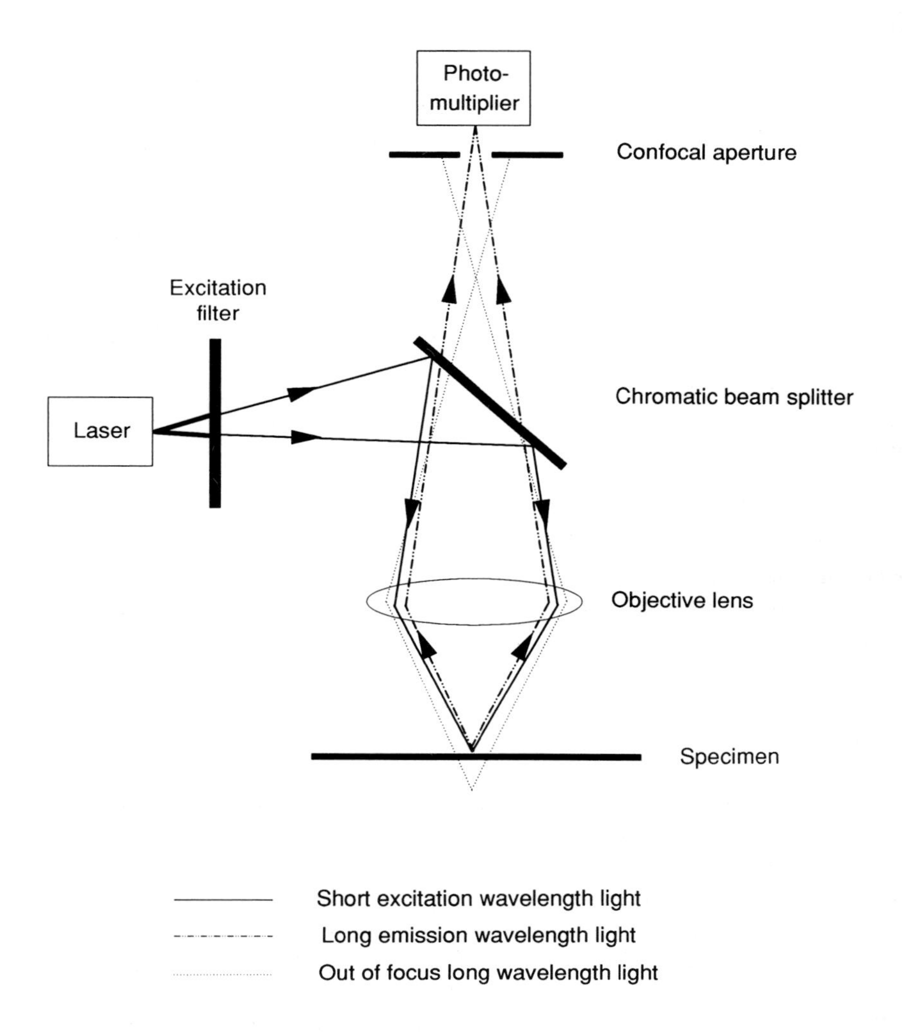

Figure 7.2: The confocal microscope.

Confocal microscopy is now used routinely by several laboratories for optical sectioning and for imaging two-dimensional material such as chromosome spreads which carry *in situ* hybridization signal. The technique can offer certain advantages.

1. The brief period of fluorochrome excitation required for producing a confocal image (usually about 2 sec) reduces the bleaching effects of prolonged excitation necessary for recording conventional epifluorescent signal.
2. The digital images recorded by the confocal microscope can be stored and recombined, enabling alignment of different images.

For chromosome analysis the ability to align accurately the probe signal to the chromosome locus and to align several probes detected by different fluorochromes on the same chromosome is useful. However, the resolution of a confocal microscope is limited by the physics of light microscopy, and for very high-resolution studies electron microscopy is needed (Section 7.3). In addition, presenting data on a high-resolution monitor records considerably less information than can be stored on fine-grained photographic emulsions used for conventional epifluorescence microscopy. In our experience the time taken to accumulate a data set using confocal microscopy is much longer than the time taken to record the data photographically.

7.2.6 Recording light microscope images

Photographic camera systems. Recording fluorescent images photographically can be difficult because extended exposure times can bleach the fluorochrome. Typical exposures on films may vary from 10 sec to 2 min depending on the film used. In addition, some camera systems are too insensitive for automatic exposure meters to work. This problem is compounded by the almost monochromatic light emitted by fluorochromes affecting exposure meters and films differently depending on their particular sensitivity at different wavelengths. However, there are many real advantages in using conventional camera systems. The choice of film can improve contrast between the signal and counterstain, long exposure times can accumulate signal and the total amount of stored information on photographic film is extremely large.

Recent developments in colour print film make it the preferred choice to record colour images with good contrast (especially in high-speed films), resolution and sensitivity. Selective enlargements can always be made from negative films and good slides from prints. Virtually all publishers will now accept colour prints or negatives. It is preferable to use as slow a film as possible (low ASA) to minimize grain size and maximize resolution, although for fluorescence photographs faster films are needed to accumulate and record the image without extended exposure times. For fluorescence photomicroscopy we routinely use Kodak Ektar 1000 or Fujicolour 400. For transmitted light microscopy (e.g. recording coloured precipitates generated by enzyme detection systems) Fujicolour 100 or Kodak Ektar 125 films give good contrast and resolution. Agfa 1000 RS is a good slide film for directly recording *in situ* hybridization signal.

Frequently images have to be recorded on colour films because the *in situ* signal can have inadequate contrast from the background in black and white. However, black and white film has the advantages of greater flexibility for enlargements, and prints are easier and cheaper to produce and publish. High-speed black and white films can often have an unacceptably large grain size. Kodak have produced films in which the silver crystals are tabular (T-max), which increases the resolution for a given speed substantially. T-max 400 is recommended for recording fluorescence images. For recording images in transmitted light, for which the speed of film is not important, small-grained films such as Agfa Ortho 25 are recommended for high-contrast images and Kodak Tech-Pan films when low contrast is required.

Digital imaging. Recording images electronically using video or low-light cameras has the advantage of image processing. Optical sections can be taken through material and out-of-focus information removed using complex algorithms. Contrast and intensity changes are easily and quickly performed and artificial colours can be assigned to sectors to enhance contrast. Different images can be overlaid and compared directly, and relative sizes of *in situ* hybridization signal calculated and compared, enabling some degree of signal quantification (Section 7.4).

One of the major uses of digital images has been to pseudocolour chromosomes after a multicolour fluorescence *in situ* hybridization approach based on combinatorial probe labelling (see Section 6.4).

Digital imaging can also be used to remove known artefacts or to improve the presentation or display of data. However, workers must guard against electronic incorporation or removal of data, which can lead to spurious results. All electronically accumulated images are recorded at low resolution compared with photographic films.

The most sensitive camera system to date is the cooled CCD (charge-coupled device) camera. This camera can record *in situ* hybridization signal which is only marginally visible by eye, because of its high efficiency in counting emitted photons over a wide spectral range. The material under investigation is located and focused by eye and then the image is accumulated and recorded over several seconds by the CCD camera and displayed on a monitor.

The problem with CCD cameras is the time required to search and record images in each field of view, especially when the signal is very weak and difficult to find on the slide. The advantage is that *in situ* hybridization signal does not have to be greatly amplified to enable visualization. This will inevitably lead to higher sensitivity and resolution.

7.3 Transmission electron microscopy (TEM)

Imaging *in situ* hybridization by electron microscopy is relatively straightforward and requires no further experience beyond conventional procedures. In general, the experimental requirements mean that the low and middle levels of resolution are appropriate, with seldom any need to image material at magnifications greater than 20 000 ×.

Block staining of embedded material with osmium or uranyl acetate is probably not possible before *in situ* hybridization because these stains can inhibit the *in situ* hybridization reaction. Consequently, contrast of the material is often very low. Increased contrast can be obtained by imaging the material at relatively low kV (e.g. 40–60 kV). However, a low kV can be damaging to material and support films and should be used cautiously. Uranyl acetate and/or lead citrate staining after *in situ* hybridization can improve the contrast of the image (Section 8.9), but care should be taken not to overstain or the *in situ* signal may be difficult to distinguish.

7.4 Quantification of signal

Ideally, knowledge of the precise location and quantity of *in situ* hybridization signal is required. As the detection sensitivity improves both requirements are increasingly being achieved. *In situ* hybridization still represents the best way to determine the number of chromosomal locations for a sequence and the location of any nucleic acid within a particular cell. However, the amount of signal is more difficult to determine and quantification techniques are still in their infancy. There is still no method to determine accurately the copy number of a nucleic acid sequence from the strength of the *in situ* hybridization signal.

Quantification of radioactive *in situ* signal can be assessed by statistically counting the number of silver grains per unit area of the tissue. This method has been used for the chromosomal assignment of genes. However, there are many factors that affect the response of the emulsion to the radiation emitted by the radioactive label (e.g. length of exposure and the particular isotope used) and so care must be taken in interpreting the results. Approaches to the quantification of radioactively labelled mRNA probes by computer-assisted image analysis are discussed by Davenport and Nunez (1990).

For the quantification of non-radioactively generated signals various methods have been used. In particular, digital imaging techniques such as the cooled CCD camera (Section 7.2.6) enables quantitative data on signal intensities (e.g. fluorochromes or density of enzyme-mediated precipitate) to be assessed. One of the major problems, however, is the large variation in the strength of the *in situ* hybridization signal even between adjacent cells. Although much effort has been put into optimizing the *in situ* hybridization procedure, this problem has still not been solved satisfactorily. An additional problem is that in many detection methods the signal is amplified. Quantification of the signal, when the amount of amplification in absolute terms is unknown, is difficult. Correction for variation in hybridization efficiency might be overcome by the incorporation of an internal standard.

Using electron microscopy, some degree of quantification has been attempted by colloidal gold detection of hybridization sites. Small grains (5 nm) of colloidal gold occur at higher density than larger grains (20–40 nm) at sites of probe hybridization (possibly due to surface charging on the grains) and are most useful for signal quantification. By counting the number of grains overlaying particular areas, a relative signal strength can be determined (Beesley, 1989).

Further reading

Albertson DG, Sherrington P, Vaudin M. (1991) Mapping nonisotopically labeled DNA probes to human chromosome bands by confocal microscopy. *Genomics* **10**, 143–150.

Baker JRJ. (1989) Autoradiography: a comprehensive overview. *RMS Microscopy Handbook* Vol. 18. Oxford University Press, New York.

Beesley JE. (1989) Colloidal gold: a new perspective for cytochemical marking. *RMS Microscopy Handbook*, Vol. 17. Oxford University Press, New York.

Cremer T, Remm B, Kharboush I, Jauch A, Wienberg J, Stelzer E, Cremer C. (1991) Non-isotopic *in situ* hybridization and digital image analysis of chromosomes in mitotic and interphase cells. *Rev. Europ. Technol. Biomed.* **13,** 50–54.

Darzynkiewicz Z. (1990) Differential staining of DNA and RNA in intact cells and isolated cell nuclei with acridine orange. pp. 285–298. In *Methods in Cell Biology*, Vol. 33, *Flow Cytometry* (eds Z Darzynkiewicz and H Crissman). Academic Press, San Diego, California.

Davenport AP, Nunez DJ. (1990) Quantification of radioactive mRNA *in situ* hybridization signals. pp. 95–111. In *In Situ Hybridization, Principles and Practice* (eds JM Polak and JO'D McGee). Oxford University Press, New York.

Johnson CV, McNeil JA, Carter KC, Lawrence JB. (1991) A simple rapid technique for precise mapping of multiple sequences in two colours using a single optical filter set. *Genet. Anal. Techniques Applications* **8,** 75–76.

Landegent JE, Jansen in de Wal N, van Ommen G-JB, Baas F, de Vijlder JJM, van Duijin P, van der Ploeg M. (1985) Chromosomal localization of a unique gene by non-autoradiographic *in situ* hybridization. *Nature* **317,** 175–177.

Narayanswami S, Hamkalo B. (1991) DNA sequence mapping using electron microscopy. *Genet. Anal. Techniq. Applic.* **8,** 14–23.

Pinkel D, Straume T, Gray JW. (1986) Cytogenetic analysis using quantitative, high sensitivity, fluorescence hybridization. *Proc. Natl. Acad. Sci. USA* **83**, 2934–2938.

Ploem JS, Tanke HJ. (1987) Introduction to fluorescence microscopy. *RMS Microscopy Handbook* Vol. 10. Oxford University Press, New York.

Rawlins DJ, Shaw PJ. (1990) Three-dimensional organization of ribosomal DNA in interphase nuclei of *Pisum sativum* by *in situ* hybridization and optical tomography. *Chromosoma* **99**, 143–151.

Shotton DM. (1989) Confocal scanning optical microscopy and its applications for biological specimens. *J. Cell Sci*. **94**, 175–206.

8 The *In Situ* Hybridization Schedule, Problems and Controls

This chapter presents schedules for *in situ* hybridization at both the light and electron microscopy level. Section 8.1 describes RNA:RNA *in situ* hybridization and Section 8.2 describes DNA:DNA *in situ* hybridization. The detection systems described are the same for both RNA:RNA and DNA:DNA hybrids; these are outlined in Section 8.3 for radioactively labelled probes, Section 8.4 for biotin-labelled probes and Section 8.5 for digoxigenin-labelled probes. For biotin- and digoxigenin-labelled probes three signal-generating systems are described: enzyme-mediated (horseradish peroxidase, Section 8.6), fluorochromes (Section 8.8) and colloidal gold (Section 8.9). A comparison of the different signal-generating systems is given in *Table 6.2*.

Details of the buffers used in the schedule and the sources of some of the reagents are found in the Appendix. Buffers are autoclaved prior to use and the expensive reagents handled aseptically.

Extreme care should be taken in handling all material. For RNA:RNA *in situ* hybridization, suitable precautions should be taken to remove contaminating RNase; these are outlined in Section 8.1. For steps in the schedule requiring small volumes (i.e. less than 300 μl) the solution is pipetted on to the slide, and plastic coverslips, made by cutting autoclavable plastic bags to a suitable size, are used. Plastic coverslips spread the fluid evenly over the material, are less damaging to specimens than glass coverslips and can easily be floated away. For electron microscopy, mesh grids, supporting sectioned material, are handled with fine forceps, immersed into small drops of solution on a glass slide and covered with a plastic coverslip. All washing steps use larger (50–100 ml) volumes to rinse slides and grids thoroughly; these are performed in Coplin jars (slides) or glass petri dishes (grids). Grids and slides should never be allowed to dry out except where specifically stated.

All reactions are carried out at room temperature unless otherwise stated.

8.1 RNA:RNA *in situ* hybridization

Precautions against RNase activity

(a) Wear disposable gloves at all times.
(b) Wherever possible use sterile disposable plasticware, which is essentially RNase free.
(c) If glassware is used, this should be baked at 180°C for at least 2 h prior to use to destroy RNase activity.
(d) General plasticware should be washed in 0.2 M NaOH for several hours and then rinsed in RNase-free water (see e).
(e) RNase-free water can be prepared by treating water with diethyl pyrocarbonate (DEPC, 0.1%), which is a strong but not complete inhibitor of RNase. The water is then autoclaved or boiled to remove all traces of the DEPC. Some workers use DEPC-treated water to make all aqueous buffers, but we find this is unnecessary and prefer autoclaved, millipore-filtered deionized water. We always use DEPC-treated water in solutions for *in vitro* transcription and the hybridization mix.

DEPC is a suspected carcinogen and should be handled with care. DEPC-treated water is prepared as follows:-

1. Make up 0.1% (v/v) DEPC solution in water (in a fume hood).
2. Stir for 2 h.
3. Autoclave the treated water for 15 min at 15 lb sq.in.$^{-1}$ (i.e. use a pressure cooker) or boil for 30 min. This step must be performed in a fume hood.

8.1.1 Pretreatments (Section 3.4)

Reagents

(a) *Pronase E*: Pronase E (Sigma, protease type XXV) is treated to remove nucleases by dissolving in water at 40 mg ml^{-1} and incubating for 4 h at 37°C. This stock solution is stored in aliquots at −20°C. Prior to use dilute the stock solution to 0.125 mg ml^{-1} in 50 mM Tris·HC1, pH 7.5 (Appendix), 5 mM EDTA.
(b) *Freshly depolymerized paraformaldehyde (4%)*: Depolymerized paraformaldehyde is prepared in a fume hood by adding 2 g of paraformaldehyde to 40 ml of 1 × PBS (Appendix), heating to 60°C for 10 min and clearing the solution with about 10 ml of 0.1 M NaOH. Adjust volume to 50 ml.

Method

1. *Pronase treatment*
   (i) Incubate slides in Pronase E for 10 min. The duration of treatment can vary with different tissues and should be determined empirically.

(ii) Stop Pronase E activity by transferring slides into 0.2% (w/v) glycine in 1 × PBS for 2 min.
(iii) Place slides in freshly depolymerized paraformaldehyde and incubate for 10 min.
(iv) Wash slides in 1 × PBS for 2 × 5 min.

2. *Acetylation*
(i) Place slides in 0.1 M triethanolamine/HCI (pH 8), which is continuously mixed using a magnetic stirrer.
(ii) Add acetic anhydride to a final concentration of 0.5% (v/v), stirring very vigorously for 10 sec. Warning: This is highly combustible, volatile and corrosive and should be handled carefully in the fume hood.
(iii) Incubate the material for 10 min, stirring gently.
(iv) Rinse the slides in 1 × PBS for 1 min.

3. *Dehydration*
(i) Rinse material in 0.85% (w/v) NaCl for 2 min.
(ii) Dehydrate in 30%, 50%, 75%, 85%, 95%, 100%, 100% ethanol for 1 min each. Slides can be stored for a few hours in ethanol vapour at 4°C before adding the probe.
(iii) Air dry immediately prior to hybridization.

8.1.2 Hybridization mix and denaturation for RNA:RNA *in situ* hybridization (Sections 5.3 and 5.4)

Probe RNA is labelled using procedures described in Chapter 4, Section 4.2.2. The hybridization mix is usually prepared immediately prior to use but can be stored at −20°C for 6 months. The hybridization mix below is for tritiated and non-isotopically labelled probes. ^{35}S-labelled probes need the addition of 20 mM dithiothreitol.

Reagents

(a) *Formamide*: Deionized, high-grade formamide is used; it must be stored at −20°C. Care must be taken with formamide because it is toxic and carcinogenic.
(b) *50% (w/v) dextran sulphate*: This is made up in DEPC-treated water, millipore (0.22 μm) filtered and stored in aliquots at −20°C.
(c) *10× salts*: 3 M NaCl, 0.1 M Tris·HCl, pH 6.8 (Appendix), 0.1 M NaH_2PO_4/ Na_2HPO_4 (pH 6.8), 50 mM EDTA in DEPC-treated water.
(d) *Blocking tRNA*: 100 mg ml^{-1} nuclease-free tRNA (Sigma, type XXI) in DEPC-treated water.
(e) *100× Denhardt's*: 2% (w/v) bovine serum albumin (fraction V, nuclease free), 2% (w/v) Ficoll 400, 2% (w/v) polyvinylpyrollidone in water.

Method

1. Prepare the hybridization mix.

Solution	Amount recommended per slide (μl)	Final concentration
100% formamide	16	50%
50% (w/v) dextran sulphate	8	10%
10× salts	4	1 ×
Blocking tRNA	0.4	1 mg ml^{-1}
100× Denhardt's	0.4	1 ×
Water	3.2	

2. Denature the RNA probe (which is in 50% (v/v) deionized formamide in water; Chapter 4, *Table 4.4*) at 80°C for 2 min.
3. Add 8 μl of denatured RNA probe to 32 μl of the hybridization mixture (above).
4. Add 40 μl of denatured hybridization mixture on to each slide and cover.
5. Place slides in a humid chamber containing tissues soaked in 2× SSC (Appendix).

8.1.3 Hybridization of RNA

Hybridize at 50°C in the humid chamber overnight.

8.1.4 Post-hybridization washes (Section 5.5)

The method described is for tritiated and non-isotopically labelled probes. For ^{35}S-labelled probes 100 mM β-mercaptoethanol should be added to the formamide washes.

Reagents

(a) *Formamide wash*: 50% (v/v) formamide in 2× SSC (Appendix).
(b) *NTE buffer*: 0.5 M NaCl, 10 mM Tris·HCl, pH 7.5 (Appendix), 1 mM EDTA.
(c) *RNase A*: 20 μg ml^{-1} RNase A (Sigma, Type 1A) in NTE buffer.

Method

1. Place slides in the formamide wash at 50°C and gently shake until the coverslips fall off (about 30 min).
2. Incubate slides for 2 × 90 min in the formamide wash at 50°C, shaking gently.
3. Wash in NTE buffer at 37°C for 2 × 5 min.
4. Incubate slides in RNase A at 37°C for 30 min.
5. Wash in NTE buffer for 2 × 5 min.

6. Incubate in the formamide wash for 90 min at 50°C, shaking gently.
7. Wash in 1× SSC (Appendix) for 5 min.
8. For the detection of radioactively labelled probes by autoradiography see Section 8.3. For the detection of biotin-labelled probes see Section 8.4 and for digoxigenin-labelled probes see Section 8.5.

8.2 DNA:DNA *in situ* hybridization

8.2.1 Pretreatments (Section 3.4)

The pretreatments are usually different for tissue sections and spread chromosome preparations.

Acetylation is rarely used except where high levels of endogenous biotin interfere with the detection of a biotinylated probe or when the slide is coated with poly-L-lysine. When necessary this step would be carried out after overnight desiccation using the method described in Section 8.1.1.

Reagents

(a) *RNase A*: Prepare stock RNase A by dissolving 10 mg ml^{-1} DNase-free RNase in 10 mM Tris·HCl, pH 7.5 (Appendix), 15 mM NaCl. Boil for 15 min and allow to cool. Store frozen in aliquots. For use dilute to 100 μg ml^{-1} RNase A in 2× SSC (Appendix).
(b) *Proteinase K reaction buffer*: 20 mM Tris·HCl, pH 8 (Appendix), 2 mM $CaCl_2$.
(c) *Proteinase K*: Prepare a 1–5 μg ml^{-1} solution of proteinase K in proteinase K reaction buffer.
(d) *Stop buffer for proteinase K*: 20 mM Tris·HCl, pH 8 (Appendix), 2 mM $CaCl_2$, 50 mM $MgCl_2$.
(e) *Pepsin solution*: Prepare a 5–10 μg ml^{-1} solution of pepsin (porcine stomach mucosa, activity 3200–4500 units per mg of protein; Sigma) in 0.01 M HCl.
(f) *Freshly depolymerized paraformaldehyde (4%)*: Depolymerized paraformaldehyde is prepared in a fume hood by adding 2 g of paraformaldehyde to 40 ml of water, heating to 60°C for 10 min and clearing the solution with about 10 ml of 0.1 M NaOH. Adjust volume to 50 ml.

Method

1. *Desiccation*
   (i) Sectioned or spread material is placed in an oven at 37°C overnight.
2. *RNase treatment (DNase-free RNase A)*
   (i) Add 200 μl of RNase, cover and incubate for 1 h at 37°C in a humid chamber.
   (ii) Wash slides in 2× SSC (Appendix) for 3 × 5 min.

3a. *Proteinase K treatment (for tissue sections)*

(i) Place material in proteinase K reaction buffer for 2 × 5 min.

(ii) Add 200 μl of proteinase K, cover and incubate for 10 min at 37°C in a humid chamber.

(iii) Stop reaction by placing material in stop buffer then wash in this buffer for 3 × 5 min.

3b. *Pepsin treatment (optional for chromosome spreads)*

(i) Place material in 0.01 M HCl for 2 min.

(ii) Add 200 μl of pepsin solution, cover and incubate for 10 min at 37°C in a humid chamber.

(iii) Stop reaction by placing in water for 2 min and wash in 2× SSC for 2 × 5 min.

4. *Prehybridization fix*

(i) Place material in freshly depolymerized paraformaldehyde and incubate for 10 min.

(ii) Wash slides in 2× SSC for 3 × 5 min.

5. *Dehydration*

(i) For cell squashes: Dehydrate for 3 min each in 70%, 90% and 100% ethanol and then air dry.

(ii) For sections: Gently blow off excess liquid. Do do not allow grids to dry out.

8.2.2 Hybridization mix and denaturation for DNA:DNA *in situ* hybridization (Sections 5.3 and 5.4)

Probe DNA is labelled using procedures described in Chapter 4. The hybridization mixture is usually prepared immediately prior to use but can be stored in the freezer for up to 6 months. A typical schedule for detecting cloned repetitive sequences is given. However, blocking DNA, SDS and probe concentration can be varied and stringency (Section 5.2) altered by adjusting the concentration of formamide and SSC.

Reagents

(a) *Formamide*: Deionized, high-grade formamide is used; it should be stored at −20°C (e.g. Sigma F7508).

(b) *50% (w/v) dextran sulphate*: This is made up in water, millipore (0.22 μm) filtered, and stored in aliquots at –20°C.

(c) *Blocking DNA*: 5 μg μl^{-1} autoclaved salmon sperm DNA (stored at −20°C) is used at concentrations of 2–250× the probe concentration.

(d) *10% (w/v) SDS (sodium dodecyl sulphate) in water*: Typically used to give a final concentration of 0.05–0.15% SDS.

Method

1. Prepare a humid chamber (this is used to prevent the material from drying out) using paper tissues soaked in 2× SSC (Appendix). Heat in a water bath or oven to raise the chamber temperature to 90°C. Leave the chamber to equilibrate.
2. Prepare the hybridization mix.

Solution	Amount recommended per slide (μl)	Final concentration
100% formamide	20	50%
50% (w/v) dextran sulphate	8	10%
20× SSC	4	2×
Probe (40 ng μl^{-1})	1	1 ng μl^{-1}
Blocking DNA	2	250 ng μl^{-1}
10% (w/v) SDS	0.5	0.125%
Water	4.5	

3. Denature hybridization mix at 70°C for 10 min then transfer to ice for 5 min.
4. Add 40 μl of the denatured hybridization mix to each slide and cover.
5. Quickly place slides into the preheated humid chamber and incubate at 90°C (water bath or oven) for 10 min. Monitor the chamber temperature near the glass slides.
6. Transfer the humid chamber with material to an incubator at 37°C. It is important to transfer the chamber quickly so that the material does not cool down too rapidly. This will allow the most similar sequences to reanneal first.

8.2.3 Hybridization of DNA

Hybridize at 37°C in the humid chamber overnight.

8.2.4 Post-hybridization washes for DNA:DNA hybrids (Section 5.5)

After hybridization it is necessary to remove non-specifically bound and weakly hybridized probe.

1. Float coverslips off in 2× SSC (Appendix) at 35–42°C.
2. Give slides a stringent wash in 20% (v/v) formamide in 0.1× SSC for 2 × 5 min at 42°C. This gives a stable sequence identity of more than 80–85%.
3. Wash slides in 2× SSC at 42°C for 3 × 3 min.
4. Take Coplin jar out of water bath and leave to cool for 5 min.

5. Wash slides in 2× SSC for 3 × 3 min.
6. For the detection of radioactively labelled probes by autoradiography see Section 8.3. For the detection of biotin-labelled probes see Section 8.4 and for digoxigenin-labelled probes see Section 8.5.

8.3 Detection of radioactively labelled probes: autoradiography (Section 6.1)

Autoradiography is carried out using a photographic emulsion (Amersham Hypercoat LM-1, Kodak NTB-2, or Ilford K5). Emulsions are applied in a darkroom fitted with a dark-red (e.g. Wratten No. 2) safelight, according to the manufacturer's instruction.

Reagents

(a) *Ethanol series*: Prepare a 30%, 60%, 80% and 95% ethanol series in 0.3 M ammonium acetate. Ammonium acetate will not fully dissolve in 95% ethanol to 0.3 M, but this does not matter.
(b) *Kodak NBT-2 emulsion*: Melt the emulsion in a water bath at 45°C for 45 min, then pour the emulsion into a beaker containing an equal volume of 0.6 M ammonium acetate at 45°C, and swirl gently to mix. Dispense aliquots into plastic slide mailers (used as dipping chambers) and leave to cool for 30 min. Store the aliquots at 4°C, wrapped in three layers of aluminium foil.
(c) *Developer*: Use Kodak D19 developer diluted 1:1 with water.
(d) *Stop solution*: 1% (v/v) glycerol, 1% (v/v) acetic acid in water.
(e) *Fixative*: 30% (w/v) sodium thiosulphate in water.

Method

1. Dehydrate the slides in the ethanol series for 1 min each, followed by 1 min in 100% ethanol, then dry.
2. *Coating slides with Kodak NBT-2 emulsion*
   (i) Warm an aliquot of emulsion to 42°C in a water bath in the darkroom, invert to mix the solution and dip a few blank slides to remove bubbles. Dip the experimental slides into the emulsion, withdraw gently at a steady rate keeping the slide vertical, and allow to drain for a few seconds, then stand in a rack to dry. When all the slides have been dipped, place the rack of slides in a light-tight box and leave to dry for at least 1 h.
   (ii) Place the slides in slide boxes containing silica gel desiccant, sealed with masking tape and triple wrapped in aluminium foil, and leave at 4°C for the required exposure time.

3. *Developing the slides*
   All the solutions (i.e. developer, stop solution and fixative) must be prechilled to 14°C before use to prevent cracking of the emulsion.
   (i) Allow the box(es) of slides to warm to room temperature for 1 h. This prevents condensation, which can reverse the latent images in the emulsion.
   (ii) In the darkroom, immerse the slides in developer for 2 min, transfer to stop solution and agitate gently for 30 sec. Fix the slides in fixative for 5 min and then wash in several changes of distilled water.

For counterstaining and mounting, see Section 8.7.

8.4 Detection of biotin-labelled probes (Sections 6.2 and 6.3)

Biotinylated probes can be detected using avidin, streptavidin or antibodies raised against biotin. Here we describe detection systems that utilize avidin conjugated to a fluorochrome or the enzyme horseradish peroxidase (HRPO) and streptavidin conjugated to colloidal gold. Conduct amplification steps only if signal strength is insufficient or if colloidal gold detection is required. See Section 8.6 for performing the DAB oxidation steps for the enzyme-mediated system, Section 8.8 for counterstaining and mounting of fluorescent-labelled conjugates or Section 8.9 for counterstaining grids to visualize colloidal gold in the electron microscope.

Reagents

(a) *Bovine serum albumin (BSA) block*: 5% (w/v) BSA in 4× SSC/Tween (Appendix).
(b) *Conjugated avidin*: Dilute appropriate conjugate in BSA block.

Detection system	Avidin conjugate	Recommended concentration for use (μg ml^{-1})
1. Fluorescence	Texas red	5
	Fluorescein	5
2. Enzyme	Horseradish peroxidase	10
3. Colloidal gold*	Unconjugated	5

* If the conjugate that is needed for visualization is colloidal gold then for reasons of steric hindrance the gold should not be incorporated at this step. Instead unconjugated avidin is used. The colloidal gold is added during the amplification steps (step ix below).

(c) *Normal goat serum block*: 5% (v/v) normal goat serum in 4× SSC/Tween (Appendix).

(d) *Biotinylated anti-avidin (raised in sheep)*: 5 μg ml^{-1} biotinylated anti-avidin in normal goat serum block.
(e) *Streptavidin gold*: 1:20 dilution of 5, 10 or 20 nm streptavidin gold (Biocell) in BSA block.

Method

1. *Detection*
   (i) Place slides in 4× SSC/Tween for 5 min.
   (ii) Add 200 μl of BSA block to each slide and apply a coverslip. Incubate for 5 min.
   (iii) Remove coverslip and drain away BSA block. Add 30 μl per slide of conjugated avidin. Replace coverslip and incubate for 1 h at 37°C in a humid chamber.
   (iv) Wash slides in 4× SSC/Tween at 37°C for 3 × 8 min.
2. *Amplification of signal*
   (v) Add 200 μl of normal goat serum block to each slide and apply a coverslip. Incubate for 5 min.
   (vi) Drain away the goat serum block and add 30 μl per slide of biotinylated anti-avidin. Replace coverslip and incubate for 1 h at 37°C in a humid chamber.
   (vii) Wash slides in 4× SSC/Tween at 37°C for 3 × 8 min.
   (viii) Incubate with BSA block as in step (ii).
   (ix) Incubate with the same labelled avidin as in step (iii). If colloidal gold detection is required use 30 μl of streptavidin gold. Incubate for 1 h at 37°C.
   (vi) Wash material in 4× SSC/Tween at 37°C for 3 × 8 min.

8.5 Detection of digoxigenin-labelled probes (Sections 6.2 and 6.3)

The digoxigenin label is detected using a signal-generating system conjugated to anti-digoxigenin. Conduct amplification steps only if signal strength is insufficient or if colloidal gold detection is required. See Section 8.6 for performing the DAB oxidation steps for the enzyme-mediated system, Section 8.8 for counterstaining and mounting for fluorescent-labelled conjugates or Section 8.9 for counterstaining grids to visualize colloidal gold in the electron microscope.

Reagents

(a) *Bovine serum albumin (BSA) block*: 5% (w/v) BSA in 4× SSC/Tween (Appendix).
(b) *Anti-digoxigenin conjugate (raised in sheep)*: Dilute appropriate conjugate in BSA block.

Detection system	Anti-digoxigenin conjugate (raised in sheep)	Recommended concentration for use
1. Fluorescence	Fluorescein Rhodamine	5 μg ml^{-1} 10 μg ml^{-1}
2. Enzyme	Peroxidase	7.5 U ml^{-1}
3. Colloidal gold*	Unconjugated	20 μg ml^{-1}

* If the conjugate that is needed for visualization is colloidal gold then for reasons of steric hindrance the gold should not be incorporated at this step. Instead the unconjugated anti-digoxigenin antibody is used. The colloidal gold is added during the amplification steps (step vi).

(c) *Normal rabbit serum block*: 5% (v/v) normal rabbit serum in 4× SSC/Tween (Appendix).
(d) *Conjugated rabbit anti-sheep IgG*: Dilute appropriate conjugate in normal rabbit serum block.

Detection system	Anti-sheep conjugate	Recommended concentration for use
1. Fluorescence	FITC Rhodamine	25 μg ml^{-1} 25 μg ml^{-1}
2. Enzyme	Horseradish peroxidase	13 μg ml^{-1}
3. Colloidal gold	Gold (10 nm)	1:20

Method

1. *Detection*
   (i) Place slides in 4× SSC/Tween for 5 min.
   (ii) Add 200 μl of BSA block to each slide and apply a coverslip. Incubate for 5 min.
   (iii) Remove coverslip and drain away BSA block. Add 30 μl per slide of anti-digoxigenin conjugate. Replace coverslip and incubate for 1 h at 37°C in a humid chamber.
   (iv) Wash slides in 4× SSC/Tween at 37°C for 3 × 8 min.

2. *Amplification*
   (v) Add 200 μl of normal rabbit serum, apply coverslip and incubate for 5 min.
   (vi) Drain away normal rabbit serum and add 30 μl of labelled anti-sheep conjugate. Incubate slides for 1 h at 37°C in a humid chamber.
   (vii) Wash in 4× SSC/Tween at 37°C for 3 × 8 min.

8.6 Enzyme-mediated system: horseradish peroxidase (Section 6.3.2)

8.6.1 DAB reaction and amplification

Horseradish peroxidase oxidizes DAB (a dangerous carcinogen) to produce a brown precipitate at the site of *in situ* hybridization. The DAB precipitate can be amplified with silver (below) and the slide can be counterstained (Section 8.7). The signal is visualized by transmitted or reflection contrast microscopy (Sections 7.2.2 and 7.2.3 respectively).

Reagents

(a) *Diaminobenzidine (DAB) detection reagent*: 5 mg of DAB in 0.5 ml water (stored frozen in aliquots) added to 9.5 ml of 50 mM Tris·HCl, pH 7.4 (Appendix).
(b) *Silver amplification, solution A (in deionized water)*: 0.2% (w/v) ammonium nitrate, 0.2% (w/v) silver nitrate, 1% (w/v) tungstosilicic acid, 0.5% (v/v) formaldehyde (diluted from stock 38 % (v/v) formaldehyde in water).
(c) *Silver amplification, solution B (in deionized water)*: 5% (w/v) Na_2CO_3.

Method

1. *DAB reaction*
   (i) Remove slides from 4× SSC/Tween and drain. Add 200 µl of DAB detection reagent, and incubate in the dark for 20 min at 4°C.
   (ii) Drain slides and add a further 200 µl of DAB detection reagent, this time with the addition of fresh hydrogen peroxide (H_2O_2; 1 µl of 30% stock solution to 2 ml of DAB detection reagent). Incubate at 4°C for 20 min.
   (iii) Stop reaction with an excess amount of water.

2. *Silver amplification of DAB precipitate*
   (i) Mix an equal volume of solution A into solution B. Immediately add 500 µl to the material, cover and monitor silver deposition under the microscope.
   (ii) Stop the reaction with water followed by 1% (v/v) acetic acid for 2 min. Then counterstain and mount (Section 8.7 for light microscopy and Section 8.9 for electron microscopy).

8.7 Counterstaining and mountants for transmitted light microscopy

8.7.1 Chromosomes

Reagents

(a) *Sörenson's buffer (pH 6.8)*: 0.03 M KH_2PO_4 and 0.03 M Na_2HPO_4.
(b) *Giemsa stain:* 4% (v/v) Giemsa (Gurr) in Sörenson's buffer. Make immediately before use. If a precipitate develops on the surface, remove it with filter paper.

Method

1. Incubate slides in Giemsa stain for 10 min.
2. Wash material with distilled water and air dry.
3. Mount in standard microscope mountant, e.g. DPX mountant (BDH) or Euparal, or temporarily mount in xylene.

8.7.2 Tissue sections

Nuclear or other counterstains should be used to provide contrast but not to detract from the impact of the *in situ* hybridization signal. Many counterstains are available, e.g. toluidine blue, haematoxylin and eosin, methyl green, neutral red and safranin, which are all described in standard histology textbooks. Material can be mounted either unstained or stained. The counterstaining procedure using toluidine blue (as in *Figure 2.7c*) is given below.

Reagent

(a) *Toluidine blue*: 0.05% (w/v) toluidine blue in water.

Method

1. Incubate slides in toluidine blue for about 1 min (determine time empirically).
2. Wash away excess stain with water.
3. Rinse in distilled water and dehydrate in 30%, 50%, 75%, 95%, 100% (v/v) ethanol in water.
4. Dip slides into Histo-Clear, then again in fresh Histo-Clear.
5. Drain the Histo-Clear briefly then add a couple of drops of DPX mountant (BDH) to each slide, apply a coverslip and gently squeeze out excess DPX/Histo-Clear. Leave the mountant to set overnight. Wash slides with *in situ* hybridization signal detected by autoradiography with detergent on the underside of the slide to remove emulsion.
6. Observe under dark- or bright-field illumination.

8.8 Counterstaining and visualization of fluorochromes by epifluorescence

8.8.1 Counterstains – DAPI and propidium iodide

DAPI (4', 6-diamidino-2-phenylindole) is a suitable counterstain because its excitation (UV) and emission (blue) wavelengths do not overlap with the fluorochromes Texas red, rhodamine or FITC. Propidium iodide (PI) can also be used as a counterstain in addition to DAPI. It fluoresces red under green excitation and also at the same wavelengths used to excite FITC *in situ* hybridization signal (green fluorescence). The maximum excitation and emission wavelengths for these DNA counterstains are given in *Table 6.3*.

Reagents

(a) *McIlvaine's buffer (pH 7.0)*:
A = 0.1 M citric acid
B = 0.2 M Na_2HPO_4
Mix 18 ml of A and 82 ml of B to make the buffer pH 7.

(b) *DAPI*: Stock solution of 100 μg ml^{-1} DAPI in water is stored at −20°C. Diluted solutions can also be stored at −20°C. For use the stock solution is diluted to 2 μg ml^{-1} DAPI in McIlvaine's buffer.

(c) *PI*: Stock solution of 100 μg ml^{-1} PI in water is stored at −20°C. Immediately prior to use the stock solution is diluted to 2.5 μg ml^{-1} in 4× SSC/Tween (Appendix).

Method

1. *DAPI*
   (i) Add 100 μl of DAPI per slide, cover and incubate for 10 min.
   (ii) Wash briefly in 4× SSC/Tween and apply antifade solution (Section 8.8.2) or, additionally, counterstain with PI (step iii).

2. *PI*
   Not to be used with Texas red, rhodamine or any other red fluorescing signal-generating system.
   (iii) Add 100 μl of PI per slide, cover and incubate for 10 min.
   (iv) Wash briefly in 4× SSC/Tween and apply antifade (Section 8.8.2).

8.8.2 Antifade

1. Apply about 50 μl of antifade solution (see Appendix for suppliers) per slide.
2. Place a thin coverslip (preferably UV transparent and of high quality) over material.

3. Gently squeeze excess antifade from the slide with filter paper.

8.8.3 Visualization

The *in situ* signal should be visualized using high-performance, UV-transmitting objectives with an immersion oil that does not autofluoresce (e.g. Leitz immersion oil). Fluorescent signal should be examined using suitable epifluorescent filter sets for FITC (Zeiss filter set = 09; Leitz filter set I2/3), Texas red (Zeiss filter set = 12 or 15; Leitz filter set N2 or N2.1) and for DAPI (Zeiss filter set = 01 or 02; Leitz filter set A; *Table 7.1*).

8.9 Counterstaining for electron microscopy

Reagents

Electron microscope stains can be purchased ready prepared. Take care when handling any heavy metal salt.

(a) *Uranyl acetate*: Make stock by saturating water with uranyl acetate and allow undissolved solid to settle (at least 1 day). Store in dark. Use cleared solution and millipore (0.22 μm) filter prior to use.
(b) *Lead citrate*: Make a solution of 0.08 M lead citrate and 0.12 M sodium citrate solution by mixing the solid in half the required water. Add 0.1 M NaOH until the solid dissolves and then add water to make the correct final volume. Store lead citrate solution in a sealed container avoiding too much carbon dioxide contamination. Carbon dioxide causes the precipitation of lead carbonate.

Method

1. Immerse each grid in a drop of lead citrate solution on a clean petri dish. Try not to breathe on the drop because of the high carbon dioxide content of exhaled breath. Incubate for 5 min.
2. Quickly transfer the grids to clean deionized or distilled water and wash thoroughly to prevent stain precipitate (at least 5 × 30 sec).
3. Immerse grids in freshly millipore-filtered saturated uranyl acetate on a clean petri dish. Incubate for 1–5 min in the dark.
4. Quickly transfer the grids to clean deionized or distilled water and wash thoroughly (5 × 30 sec).
5. Drain excess liquid, air dry and examine by transmission electron microscopy.

8.10 Troubleshooting

8.10.1 Poor signal

The worst possible outcome of an *in situ* hybridization experiment is to get no signal. When this happens it is difficult to determine the source of the problem. When there is some signal, however faint or 'dirty', improving the signal is a matter of adjusting the parameters to reduce background, non-specific hybridization or cross-hybridization or to increase signal strength. It is therefore essential that the initial experimental parameters are chosen to obtain an *in situ* signal alongside suitable controls. In particular, the probe and detection reagents should be tested prior to use and hybridization and washing stringency on the first run should be low.

Probe. To test the probe, a small quantity is bound to nitrocellulose or charged nylon membrane and tested for incorporated label (i.e. as a dot blot). Radioactively labelled probes can be tested by scintillation counting. For non-radioactively labelled probes (*Table 4.9*) the result gives a semiquantitative measure of the label incorporation into the probe. When testing non-radioactive probes ideally the same detection reagents (e.g. antibodies) are used as for the *in situ* hybridization experiment to verify the effectiveness of the detection reagents.

The detection reagents are further tested during the *in situ* experiment by running, in parallel, a control slide using a reliable probe (e.g. high copy number) on material known to contain the target sequence.

Stringency. The *in situ* hybridization experiment should initially be conducted at low stringency (ideally 70–80% probe and target similarity) to encourage the binding of probe to the target sequence *in situ*. The signal strength, levels of background and cross-hybridization to non-target sequences can be assessed at this stringency. Increasing the stringency of hybridization and post-hybridization washing has the effect of removing non-specific hybridization and weakly bound probe, but can also reduce the signal strength. When using non-radioactively labelled probes the signal can be amplified to compensate for this.

DNA denaturation conditions. When there is no *in situ* hybridization signal, assuming the probe labelling and detection systems are working suitably, this is probably caused by a problem in DNA denaturation. Chromosomal DNA has a narrow window at which optimal denaturation occurs (Section 5.3). The denaturation conditions depend on the type of sequence and the extent to which it is protected by DNA-associated proteins.

If the denaturation conditions are not adequate, then the target DNA will not separate into single strands and the probe DNA will be unable to

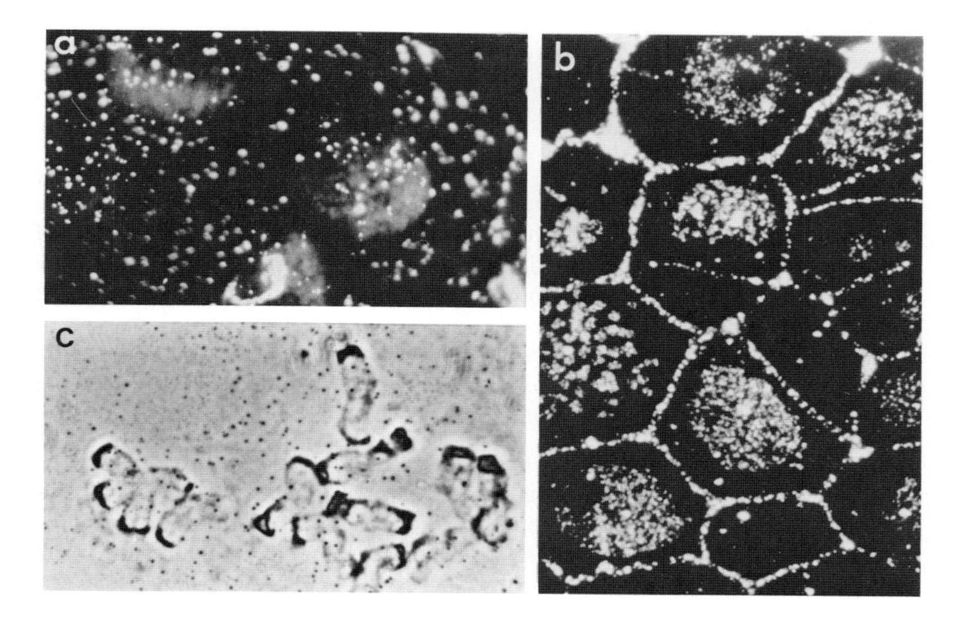

Figure 8.1: Troubleshooting – problems with *in situ* hybridization. (a) Digoxigenin-labelled probe hybridized to root tip nuclei and detected by fluorescein. The signal appears as very bright specks in several focal planes. This suggests either that the labelled probe has precipitated and not hybridized *in situ* at all or that the material is covered with a thick layer of cytoplasm or dirt which has bound the probe and/or detection reagents non-specifically. (b) Biotin-labelled probe hybridized to a root tip section and the sites of probe hybridization detected by Texas red fluorescence. The *in situ* hybridization signal primarily occurs on the nuclei and cell walls. The probe would be expected to hybridize to regions within the nucleus only. Probe hybridization to non-target sequence and nucleic acid-binding molecules can be greatly reduced by incorporating into the probe mixture an excess amount of an unlabelled nucleic acid which is not related to the probe. Weakly hybridized probe sites can be removed by increasing the washing stringency. (c) Biotin-labelled probe hybridized to chromosomes of rye and detected by DAB. Although the *in situ* signal is localized to the ends of the chromosomes (as expected), the signal is obscured by background dots of DAB precipitate. This can occur if the detection reagents are poor, the hydrogen peroxide used in the precipitation of the DAB is at too high a concentration or the material contains endogenous peroxidases. The chromosomes are also 'ghost-like' and ballooned, which suggests over-denaturation and the loss of chromosomal DNA.

hybridize to the sequence *in situ*. Conversely, if the material is over-denatured, excessive DNA loss from the material will occur. Overdenaturation is usually clear to see after *in situ* hybridization as chromosomes appear ragged and ghost-like (*Figure 8.1c*). If the counterstaining of the chromosomes is weak (e.g. DAPI; Section 8.8.1) and there is little or no *in situ* hybridization signal, then the results, together, are suggestive of over-denaturation of the chromosomal DNA.

8.10.2 Too much background/non-specific hybridization

Increasing stringency is one of the many ways to reduce background hybridization. Others include material pretreatment, increased washing steps or the addition of unlabelled nucleic acids to the probe mixture. All the methods can be used separately or in combination.

Increase pretreatments (Section 3.4). RNase digestion, to remove RNA, is an important pretreatment when detecting DNA sequences (especially if the sequence is highly expressed) because RNA can contribute significantly to background signal. Pepsin or proteinase K pretreatments can be increased to remove cytoplasmic proteins.

Acetylation of amino groups can reduce non-specific electrostatic binding of the probe. Acetylation can be used to reduce the levels of endogenous biotin, which may be important if probes are labelled with biotin.

Increase washing steps. Background can be reduced by increasing the number and duration of washing steps (e.g. post-hybridization washes and washing during the detection steps) or by incorporating harsher detergents (e.g. Triton X-100). Care must be taken not to damage or lose the material with overzealous washing conditions.

Removal of 'sticky' non-specific binding sites. If background signal is not removed by increasing stringency, pretreatments and/or washing, then the problem may relate to the non-specific binding of the probe or detection reagents to endogenous components in the system. This can be tested by conducting an *in situ* hybridization experiment without any labelled probe at all and examining the background labelling.

The effect of probe-binding molecules, which can contribute significantly to background labelling (*Figure 8.1b*), can be reduced by adding an excess of unlabelled DNA (e.g. salmon sperm DNA for DNA:DNA hybridization) or tRNA (for RNA:RNA hybridization) to the probe hybridization mix.

Detection systems for non-isotopic labels use antibodies or (strept)avidin (Section 6.2). Non-specific binding of these molecules to the material can contribute to background labelling and can be reduced using a high concentration of a protein solution (e.g. BSA or normal serum) which may preferentially bind to these 'sticky' sites. In addition, reducing the concentration of detection reagents may reduce some background.

Changing of probe label. If the above methods fail to eliminate the background, then another likely cause is endogenous label in the cytoplasm of the cell. In particular, some cells have high levels of endogenous biotin (e.g. some animal and plant tissues), which may be detected if a biotinylated probe is used. In this case it is worth changing the probe label (e.g. to digoxigenin).

8.10.3 Patchiness

Almost any slide with *in situ* hybridization shows patchiness of signal. The patchiness probably arises during tissue preparation, especially during tissue squashing or spreading. In particular, cytoplasmic debris around chromosomes will inhibit *in situ* hybridization (e.g. *Figure 8.1a*). This could be because the DNA is heavily protected by the cytoplasm, preventing probe and detection reagent access, DNA denaturation, or a combination. Therefore cell spread preparations should be as clean as possible with little interfering cytoplasm.

Patchiness of signal can also result if the reagents have not been well mixed or if air bubbles form under the coverslips during the incubations steps.

8.11 Controls

Controls should be incorporated with all *in situ* hybridization experiments. At all times Northern and Southern blots can be carried out to determine the specificity of probe hybridization to target sequences. A number of other controls can also be performed. These are outlined below.

8.11.1 Determining the level of non-specific hybridization

Conduct the experiment either without adding any probe, adding unlabelled (cold) probe or using an unrelated probe known to be absent from the material. For RNA:RNA *in situ* hybridization experiments the sense strand, synthesized during *in vitro* transcription (Section 4.2.2), is a particularly useful control which should not hybridize (*Figure 2.7e*).

For DNA:DNA *in situ* hybridization digesting the material with DNase will remove the target sequence. However, this control is rarely, if ever, performed. The equivalent experiment for RNA:RNA *in situ* hybridization, digesting material with RNase, is not recommended because RNase activity is difficult to remove and will potentially degrade the probe.

Conduct the experiment on material that is known not to contain the target sequence of interest.

8.11.2 Determining the specificity of probe hybridization

Conduct several experiments at different stringencies to determine the effect on the strength and specificity of the hybridization signal (Section 5.2, *Figure 2.1h*). Another control used with RNA:RNA *in situ* hybridization experiments uses different regions of a cDNA separately as a probe. Each region should give the same result.

8.11.3 Positive controls

Positive controls should also be carried out to demonstrate that denaturation, hybridization and detection reagents have all worked properly. For DNA:DNA *in situ* hybridization this should involve the detection of a reliably labelled, highly repeated sequence.

For RNA:RNA *in situ* hybridization a positive control, such as the detection of ribosomal RNA, which should be detected in all cells, can be used to demonstrate RNA retention and the accessibility of target sequences in all cell types (*Figure 2.7f*). This is particularly important in experiments designed to determine differential expression of a gene. Another control can be conducted by using a mutant (if available) which is known not to be expressing the gene of interest.

8.12 Safety

Many chemicals used for *in situ* hybridization are dangerous and include toxic, allergenic, carcinogenic and teratogenic substances. All chemicals should have a proper safety assessment before being used. These can be obtained from the manufacturer, and a laboratory new to any reagent should be aware of all risks and potential risks.

At all times, good laboratory practice should be used (e.g. a laboratory coat and gloves should be worn and dangerous chemicals and vapours should be handled in a fume hood). When weighing dangerous powders in a fume hood (e.g. paraformaldehyde, fluorochromes, diaminobenzidine) a sample should be weighed to approximately the desired quantity and diluted to the correct concentration by altering the liquid content. Altering the solid content to add to a fixed volume increases the amount of handling.

Procedures involving dangerous, vaporous solutions, especially those that are heated (e.g. formaldehyde, formamide), should be carried out in a fume hood. The dangers of radioactive nucleotides are minimized using proper laboratory practice in a laboratory commissioned for handling radiation.

9 The Future of *In Situ* Hybridization

In situ hybridization will increasingly address important biological problems. Already it has made a major impact in cell biology and cytogenetics (e.g. the Human Genome Project) and its applications will increase. The diversity of approaches presented here (e.g. EM level detection and multiple labelling) will be more widely used, particularly in the detection of mRNA *in situ*, than has been reported to date. Flow cytogenetics following *in situ* hybridization of cells and chromosomes in suspension may become an important research tool in cell biology and medical diagnosis. This will enable the analysis of millions of chromosomes or cells, giving statistically meaningful and quantifiable results of high accuracy. The probes continuously produced from this new approach will provide their own application. Adaptations of methods utilizing PCR to material studied *in situ* are making an impact, and these methods are likely to be used increasingly to localize sequences which are in low abundance, difficult to clone or defined by only a few bases (e.g. microsatellite sequences).

Advances are needed to improve routine screening. Already software exists which can identify and find specific structures (e.g. chromosomes) on glass slides, thereby preventing hours of searching. The increased availability of preprepared probes and *in situ* hybridization kits will assist in the routine use of *in situ* hybridization in diagnostic or screening laboratories.

The development of new, more versatile, signal detection systems and improved multiple labelling is occurring all the time. This is already enabling sequences to be used as markers (like a lambda ladder on a Southern hybridization experiment), while other target sequences are simultaneously and accurately localized. With increasing knowledge of the effects of stringency, *in situ* hybridization could become extremely informative about the sequences within the probe, enabling characterization, at least in part, of the probe itself.

In situ hybridization is already giving fundamental insights into how cellular events interrelate and how sequences are organized, transcribed, spliced and transported. The data will continue to influence many fields of research and will inevitably lead to new, unforeseeable, applications.

Appendix

Suppliers of reagents

Many reagents are obtainable from chemical companies. The ones below are those that the authors use regularly.

Amersham International plc, Northern Europe Region, Amersham, Little Chalfont, Buckinghamshire HP7 9NA, UK.

Boehringer Mannheim GmbH, Biochemica, PO Box 310 120, D-6800 Mannheim 31, Germany; or Boehringer Mannheim UK (Diagnostics and Biochemicals) Ltd, Bell Lane, Lewes, Sussex BN7 1LG, UK.

Enzo Diagnostics Inc., 325 Hudson Street, New York, NY 10013, USA; or Universal Biologicals Ltd, 12–14 St Ann's Crescent, London SW18 2LS, UK.

Gibco BRL, Life Technologies Ltd, 8400 Helgerman Court, Gaithersburg, MD 20877, USA; or Life Technologies Ltd, Trident House, PO Box 35, Renfrew Road, Paisley PA3 4EF, Renfrewshire, Scotland, UK.

Pharmacia LKB Biotechnology AB, Björkgatan 30, S-75182 Uppsala, Sweden; or Pharmacia Ltd, Pharmacia LKB Biotechnology Div., Midsummer Boulevard, Central Milton Keynes, Buckinghamshire MK9 3HP, UK.

Promega Corporation, 2800 Woods Hollow Road, Madison, WI 53711–5399, USA; or Promega Ltd, Epsilon House, Enterprise Road, Chilworth Research Centre, Southampton SO1 7NS, UK.

Sigma Chemical Company Ltd, PO Box 14508, St Louis, MO 63178, USA; or Sigma Chemical Company Ltd, Fancy Road, Poole, Dorset BH17 7NH, UK.

Stratagene Cloning Systems, 11099 North Torrey Pines Road, La Jolla, CA 92037, USA; or Stratagene Ltd, 140 Cambridge Innovation Centre, Cambridge Science Park, Milton Road, Cambridge CB4 4GF, UK.

Vector Laboratories Inc., 30 Ingold Road, Burlingame, CA 94010, USA; or Vector Laboratories, 16 Wulfric Square, Bretton, Peterborough, Cambridgeshire PE3 8RF, UK.

Specific products were obtained as follows:

Digestion enzymes
Calbiochem Corporation, PO Box 12087, San Diego, CA 92112–4180, USA; or

Calbiochem Novabiochem (UK) Ltd, 3 Heathcoat Building, Highfields Science Park, University Boulevard, Nottingham, NG7 1BR, UK; Yakult Pharmaceutical Ind. Co. Ltd, 1–19 Higashi Shinbashi Minato-Ku, Tokyo 105 Japan.

Biotinylated nucleotides
Enzo Diagnostics Inc: Sigma; Gibco BRL (see above).

Fluorochrome conjugated nucleotides
Amersham International plc; Boehringer Mannheim (see above).

Radioactively labelled nucleotides
New England Nuclear, Du Pont (UK) Ltd, Biotechnology Systems Division, Wedgewood Way, Stevenage, Hertfordshire, SG1 4QN UK; Amersham International plc (see above).

pGem Z, riboprobe kits
Promega (see above).

pBluescript II phagemid vector
Stratagene (see above).

DNA labelling enzymes, nucleotides and kits
Enzo Diagnostics Inc; Gibco BRL; Boehringer Mannheim; Amersham (see above).

Chemiprobe kit
FMC BioProducts, 5 Maple Street, Rockland, ME 04841, USA; Flowgen Instruments Ltd, Broad Oak Enterprise Village, Broad Oak Road, Sittingbourne, ME9 8AQ, UK.

Digoxigenin and its detection reagents
Boehringer Mannheim (see above).

Avidin conjugates (fluorochromes and enzymes), biotinylated anti-avidin, normal sera, Vectabond, Vector red
Vector Laboratories (see above).

Gold conjugates
British Biocell International, Ty-Glas Avenue, Llanishen, Cardiff, UK; Amersham and Boehringer Mannheim (see above).

Secondary antibodies
DAKO Ltd, 16 Manor Courtyard, Hughenden Avenue, High Wycombe, Buckinghamshire, HP13 5RE, UK; Vector Laboratories (see above).

Antifade
Vector Laboratories (see above).

Glutaraldehyde, EM reagents including stains, grids and support film reagents
Taab Laboratories Ltd, 40 Grovelands Road, Reading, Berkshire, UK; Agar Scientific Ltd, 66a Cambridge Road, Stansted, Essex, CM24 8DA, UK.

LR White acrylic resin (medium grade)
The London Resin Co., Ltd, PO Box 34, Basingstoke, Hampshire, RG21 2NW, UK; Taab Laboratories (see Glutaraldehyde).

Programmable temperature controller
Cambio, 34 Millington Road, Cambridge, CB3 9HP, UK.

Embedding medium for frozen tissue sections
(*Tissue-Tek*): Agar Scientific Ltd, 66a Cambridge Road, Stansted, Essex, CM24 8DA, UK.

Companies supplying *in situ* hybridization kits

Amersham International plc (see above).

R&D Systems Europe Limited, 4–10 The Quadrant, Barton Lane, Abingdon, Oxon, OX14 3YS, UK.

Cambio, 34 Millington Road, Cambridge, CB3 3HP, UK.

Genetrix, 6401 East Thomas Road, Scottsdale, AZ 85251, USA.

Gibco BRL, Life Technologies Ltd (see above).

Imagenetics, PO Box 3011, Naperville, IL 60566–7011, USA.

Integrated Genetics IG Laboratories Inc., One Mountain Road, Framingham, MA 01701, USA.

Oncor, 209 Perry Parkway, Gaithersburg, MD 20877, USA.

Buffers

Commonly used buffers are:

1. 20× SSC pH 7.0
   3 M NaCl
   0.3 M sodium citrate.
2. 4× SSC/Tween
   4× SSC plus 0.2% (v/v) Tween 20.
3. 1× PBS (pH 7.4): available as a powder from Sigma. To make:
   120 mM NaCl
   7 mM Na_2HPO_4
   3 mM NaH_2PO_4
   2.7 mM KCl.
4. Tris·HCl
   Make up to required molarity of Trizma base with water, adjust pH with 1 M HCl and make up to final volume with water.

5. 100× TE pH 8.0
   1 M Tris·HCl, pH 8.0
   0.1 M EDTA, pH 8.0.

Index